WELCOME TO YOUR CHILD'S BRAIN

BY THE SAME AUTHORS

Welcome to Your Brain: Why You Lose Your Car Keys but Never Forget How to Drive and Other Puzzles of Everyday Life

How the Mind Grows from Conception to College

WELCOME TO YOUR CHILD'S BRAIN

SANDRA AAMODT, PH.D.
AND SAM WANG, PH.D.

Foreword by Ellen Galinsky

BLOOMSBURY
NEW YORK • LONDON • NEW DELHI • SYDNEY

Art credits:
Illustrations throughout by Lisa Haney, except as noted. Pages 12, 14, 19, 35, 92, 95, 102, 106, 114, 126, 157, and 223 by Patrick Lane. Pages 44, 116, and 203 by Sam Wang. Page 64 by Ken Britten. Page 86 ©iStockphoto/ Daniel Rodriguez, modified by Ken Britten. Page 122 courtesy of Gordon Burghardt. Page 130 courtesy of Roger Tsien and Le Prix Nobel 2008.

Published by Bloomsbury USA, New York

All papers used by Bloomsbury USA are natural, recyclable products made from wood grown in well-managed forests. The manufacturing processes conform to the environmental regulations of the country of origin.

Library of Congress Cataloging-in-Publication Data has been applied for.

ISBN: 978-1-59691-649-4 (hardcover)

First published by Bloomsbury USA in 2011
This paperback edition published in 2012

Paperback ISBN: 978-1-60819-933-4

1 3 5 7 9 10 8 6 4 2

Designed by Elizabeth Van Itallie
Printed in the U.S.A. by Quad/Graphics, Fairfield, Pennsylvania

From Sandra, to my parents, of course

From Sam, to Vita, Becca, Rita, and Dad

CONTENTS

PART 6 – YOUR CHILD'S BRAIN AT SCHOOL

PART 7 – BUMPS IN THE ROAD

FOREWORD

In 1994, the Carnegie Corporation of New York released a seminal report, "Starting Points: Meeting the Needs of Our Youngest Children." Although it included only a few sentences about the brain development of young children, the ensuing media attention was all about brain development—all the time and everywhere.

Two years later, prompted by this unprecedented attention, the organization I lead, the Families and Work Institute—as well as the Carnegie Corporation, the Harris Foundation, and others—convened a conference at the University of Chicago titled "Brain Development of Young Children: New Frontiers for Research, Policy, and Practice." We brought together 150 neuroscientists with researchers who study children's social, emotional, and intellectual development, and asked: Are the findings from neuroscience similar to or different from the findings from other child development studies? How can these findings promote the healthy development and learning of all young children?

When the report from our conference was released, public interest and media attention were once again off the charts. In the months that followed, a public engagement campaign on children from birth to age three was launched, a White House conference was held, a *Newsweek* magazine special edition was published, sessions at the National Governors Association were held, and special series on the brain development of young children were produced on the *Today* show and *Good Morning America.*

In the more than a decade and a half since, public interest in the brain has continued to increase. But along with that interest, a great deal of misinformation has been promulgated. Toy creators, media developers, and marketers have stepped in "where angels fear to tread," producing programs and creating materials with claims of building bigger and better brains.

Welcome to Your Child's Brain, by Sandra Aamodt and Sam Wang, is an important antidote to all of this misinformation. As befits authors who have forty combined years covering neuroscience—Aamodt as the editor of *Nature Neuroscience* and as a science writer, and Wang as a professor and researcher at Princeton—this book is encyclopedic in its coverage. I expect it to become an invaluable reference book to help families and educators explore children's development, whenever questions and issues arise. Importantly, it can counter some of the myths that have arisen as a result of almost two decades of misinterpretations and misinformation. For example:

Does listening to music make children smarter? The authors state that there is no scientific evidence for this idea. However, they note that there can be some benefit from playing music because it can help children learn to focus.

Can we multitask? The authors write: "The brain cannot concentrate on more than one thing at a time." They go on to say that "the cost of chronic multitasking may include diminished performance when single-tasking."

Can babies learn from the media? The authors report: "No reliable research shows that TV watching has any benefits for babies." For older children, the benefits depend on what they are watching and how much time they are spending doing so.

Does birth order affect personality? The authors write: "Sorry, firstborn readers, but there is little credible evidence that birth order influences personality."

Having spent the past ten years reviewing many of the same studies the authors review for my own book, *Mind in the Making*, I wholeheartedly agree with their conclusion that the "best gift" you can give your children is self-control. They note that self-control and other executive functions of the brain (like working memory, flexible thinking, and resisting the temptation to go on automatic) "contribute to the development of [children's] most important basic brain function: the ability to control their own behavior in order to reach a goal." Noting that self-control underlies so many capacities, from socialization to schoolwork, they state that "preschool children's ability to resist temptation is a much better predictor of eventual academic success than their IQ scores."

It was a very similar realization from my own review of the research that led me to postulate seven life skills to help children succeed emotionally, socially, and intellectually. All of these life skills are based on executive functions of the brain. Importantly, Aamodt and Wang bust another myth when they report that we promote self-control not by making children sit still, chained to their desks, but by encouraging them to play. They write, "It probably doesn't matter exactly what excites your children; as long as they are intensely engaged by an activity and concentrate on it, they will be improving their ability to self-regulate and thus their prospects for the future."

Welcome to Your Child's Brain is a very welcome addition!

—**Ellen Galinsky**, author of *Mind in the Making: The Seven Essential Life Skills Every Child Needs* and president of the Families and Work Institute

Quiz

HOW WELL DO YOU KNOW YOUR CHILD'S BRAIN?

---

1) Which of the following is a good way to get your child to eat his spinach?

a. Cover the spinach in melted cheese

b. Start the meal with a few bites of dessert

c. Feed him with soy-based formula as an infant

d. All of the above

e. None of the above

2) Which action by a pregnant woman poses the greatest risk to her baby?

a. Drinking a beer one evening

b. Fleeing from a hurricane

c. Eating sushi for dinner

d. Flying in an airplane

e. Walking three miles

3) What fraction of the calories eaten by a five-year-old go to power her brain?

a. One tenth

b. One quarter

c. One half

d. Two thirds

e. Nearly all

4) How do your child's genes and his environment interact during development?

a. His genes influence his sensitivity to environmental features

b. His environment influences the expression of his genes

c. His genes influence how you care for him

d. His genes and his environment are inseparably entangled

e. All of the above

5) Which of the following increases a baby's intelligence?

a. Breast-feeding during infancy

b. Eating fish during pregnancy

c. Listening to Mozart

d. All of the above

e. None of the above

6) If you cover a doll, then remove the drape, what will a baby be most surprised to see?

a. Two dolls

b. A toy car

c. An upside-down doll

d. A stuffed octopus

e. A cheeseburger

7) Which of the following activities is likely to improve a child's school performance?

a. Studying with a friend

b. Listening to music while studying

c. Taking breaks from studying to play video games

d. All of the above

e. None of the above

8) What kind of dream experience is not yet within the capacity of a three-year-old?

a. Seeing a dog standing around

b. Playing with toys

c. Sleeping in the bathtub
d. Watching tropical fish
e. Looking at an empty room

9) What accelerates the ability to understand what other people are thinking?

a. Learning a second language
b. Having an older sibling
c. Parents who talk about emotions
d. All of the above
e. None of the above

10) Which of these activities reduces a child's risk of nearsightedness?

a. Eating fish
b. Playing outside
c. Learning a musical instrument
d. Getting enough sleep
e. Resting her eyes

11) Which categories can an infant distinguish?

a. Male faces from female faces
b. Major chords from dissonant chords
c. The mother's language from a foreign language
d. All of the above
e. None of the above

12) Which of the following interventions has the largest effect?

a. Ballet lessons improve gender identity
b. Etiquette lessons improve empathy
c. Moral lessons improve behavior
d. Music lessons improve math ability
e. Drama lessons improve social adjustment

13) What increases a child's likelihood of becoming autistic?

a. Being born prematurely

b. Having an unresponsive mother

c. Watching too much TV

d. Receiving vaccinations

e. Having older siblings

14) How does Ritalin improve focus in kids with attention-deficit/ hyperactivity disorder?

a. By altering the structure of brain circuits

b. By sedating the child slightly

c. By activating the same brain cells as cocaine and amphetamine

d. All of the above

e. None of the above

15) Which of the following activities improves self-control?

a. Pretending to be a fireman

b. Being breast-fed

c. Watching baby videos

d. Sleeping with parents

e. Having cross-gender friendships

16) Brain imaging can do which of the following?

a. Diagnose behavior disorders

b. Predict reading and math ability

c. Determine when a criminal is lying

d. All of the above

e. None of the above

17) Which of these experiences is most associated with future reading difficulties?

a. Being deprived of children's books as a toddler

b. Writing letters backward at the age of four

c. Having difficulty sitting still

d. Speaking two languages

e. Not hearing enough children's music early in life

18) Which of the following is most likely to improve a shy child's future life?

a. Having parents who read science books
b. Playing with toy trucks
c. Growing up in China
d. All of the above
e. None of the above

19) Where did the disciplinary concept of time-out originate?

a. The rules of organized sports
b. Frustrated parents
c. Computer engineering
d. Military slang
e. Studies of learning in laboratory animals

20) How do children who learn a second language compare with monolingual children?

a. Better at self-control
b. Better at taking the perspective of other people
c. Experience decline in intellectual sharpness later in adulthood
d. All of the above
e. None of the above

Key: 1) d, 2) b, 3) c, 4) e, 5) b, 6) a, 7) c, 8) b, 9) d, 10) b, 11) d, 12) e, 13) a, 14) c, 15) a, 16) e, 17) a, 18) c, 19) e, 20) d

Introduction

THE BRAIN THAT BUILDS ITSELF

Moms and dads ask a lot of questions. My son says video games make him smarter—is that possible? How essential is it to breastfeed the baby? Is it okay to eat fish during pregnancy? Are vaccines safe for kids? My preschooler is writing her *R*s backward—is she dyslexic? And why can't I drag my teenager out of bed?

Call us geeks, but when we hear such concerns, we think about neuroscience. All of these questions involve the brain and how it develops. Childhood is a period of dramatic brain growth and behavioral change, and parents have a front-row seat. If you find this process as fascinating as we do, or if you're simply looking for some answers, this book is for you.

We cover the entire period from conception to college—because brain development goes on a lot longer than the first three years, where many other books stop. The growth and maturation of a child's brain is an intricate process taking decades, in which the brain grows and adapts to the surrounding world. The job won't be entirely finished until your child is in college. So whether your child is an infant, a toddler, or a teen, read on.

Between us, we have over forty years of experience as neuroscientists. Sandra started in the laboratory doing research on brain development and **plasticity*** and went on to edit one of the leading journals in neuroscience. She has read thousands of papers, many of them reporting pioneering discoveries. Her critical eye comes from having a view of the field that is both broad and deep. She knows when a result is sound and when it is fishy.

Sam is a professor and researcher at Princeton University. He has been publishing original research and teaching students for over twenty years. His own

*Bold terms are defined in the glossary. See pp. 267–273.

research concerns how the brain processes information and learns—and how this process can go wrong in early life.

Sam is also a dad. Before his daughter came along, he used to talk about what we called *cocktail party neuroscience*. Life changed for him, so now it's *preschool potluck neuroscience*. At these parties, parents and teachers ask lots of fun questions, but sometimes he's noticed a touch of anxiety as well.

Your questions sent us to the library. Together we scoured the technical literature, studying many hundreds of papers in neuroscience, psychology, medicine, and epidemiology. We synthesized this vast literature into our best interpretation of what is known about children's brains. This book is the result of all that research. In it, we explain the science, debunk myths, and include practical tips for you as parents.

Here's our first instruction: **take a deep breath and relax**. Really. The things you're worrying about are much smaller factors in your child's well-being than you might imagine. Many modern parents believe that children's personality and adult behavior are shaped mainly by parenting—but research paints a very different picture.

There is a simple way to summarize much of the research on the neuroscience of child development: children grow like dandelions. In Sweden, the term *maskrosbarn* (**dandelion child**) is used to describe children who seem to flourish regardless of their circumstances. Psychological studies suggest that such children are relatively common (at least when raised by "good enough" parents who do not abuse or neglect them). From an evolutionary perspective, this makes sense; children who can make do with whatever time and attention their parents can spare are more likely to survive and pass along their **genes** under tough conditions. For many brain functions, from temperament to language to intelligence, the vast majority of children are dandelions.

Here's our first instruction: take a deep breath and relax. Really.

The developing brain has been shaped by thousands of generations of evolution to become the most sophisticated information-processing machine on Earth. And, even more amazingly, it builds itself. For instance, you do not need to teach your children to notice—and eventually produce—human speech. Your baby son or daughter knows, very early on, that the noises you make have more meaning than other sounds. So even if you never give your children a speech lesson, they are highly unlikely to start imitating the air conditioner or the family cat. At least not convincingly.

Children are not passive recipients of parenting or schooling, but active participants in every aspect of their own development. From birth, their brains are prepared to seek out and make use of experiences that suit their individual needs and preferences. For this reason, brain development requires no special equipment or training, and most children find a way to grow in whatever conditions the world has to offer them.

If children are so adaptable and smart, why can't they start using their brains for high-powered activities right away? In large part it's because the development process tunes each individual's brain to the characteristics of a particular environment. This is one reason that people can live successfully all over the world. Genes provide the blueprints for your child's individuality, but the plans are certain to be modified during construction depending on local conditions—not only your actions as parents, but also your child's culture, neighborhood, teachers, and peers. This matching process is automatic, with some support from you along the way. All this leads us to the major theme of this book: **your child's brain raises itself**.

In a few circumstances, extra help is necessary. Things can go wrong if the genetic program has a flaw or if environmental conditions are very difficult, as happens in poverty or war. Modern life has also created some new challenges. Brain development can get into trouble when our modified environment fails to play nicely with our ancient genetic heritage. For these cases, we tell you how to give your child that extra boost.

We organized the book around seven scientific principles that will help you understand how your child's brain grows and changes along the path to adulthood.

- **Part 1, Meet Your Child's Brain.** This section is an introduction to your child's brain and how it works. In particular, we talk about how innate predispositions for interacting with the outside world initiate a two-way conversation between genes and environment that shapes neural development throughout childhood.

- **Part 2, Growing Through a Stage.** The brain goes through periods when it builds upon earlier foundations and is exceptionally sensitive to certain types of information. This section describes the experiences that your child's brain uses to shape the development of sleeping, walking, and talking.

- **Part 3, Start Making Sense.** Much of neural development relies on experiences that are easily available to almost any child. As parents, you get a free ride on this process; simply sit back and watch your child's senses tune themselves to the world.

- **Part 4, The Serious Business of Play.** One of the major ways that children adapt to their circumstances is through play. From preschool through adolescence, play is practice for adult life and helps to develop some of the brain's most important functions.

• **Part 5, Your Child as an Individual.** Distinctive features of the genetic program make your baby a unique person from the start. Here we explain how your child's individual emotional and social characteristics grow and respond to the surrounding world.

• **Part 6, Your Child's Brain at School.** Most of the evolutionary history of our species happened before there were books, violins, or calculus—not to mention Facebook. We tell you how the flexibility of your child's brain allows her to handle abstract concepts that our ancestors never imagined.

• **Part 7, Bumps in the Road.** All environments present challenges to the developing brain. Most children can get what they need to grow, like dandelions, but a few are more delicate flowers needing extra care or attention. We explore what you can do to help your child if anything goes wrong.

Feel free to dip in anywhere that interests you. Headings indicate the age range that is the focus of each chapter, so that you can easily find out whether we have something to say about your child's brain, however old he or she is right now. As you can see, we've got a lot of ground to cover, so let's get started.

PART ONE

MEET YOUR CHILD'S BRAIN

THE FIVE HIDDEN TALENTS OF YOUR BABY'S BRAIN

IN THE BEGINNING: PRENATAL DEVELOPMENT

BABY, YOU WERE BORN TO LEARN

BEYOND NATURE VERSUS NURTURE

Chapter 1

THE FIVE HIDDEN TALENTS OF YOUR BABY'S BRAIN

AGES: BIRTH TO ONE YEAR

Your baby is smarter than he or she lets on. For generations, the slow development of **motor** systems led psychologists to believe that babies had very simple mental lives. In a baby who has not worked out how to walk or talk, mental capacities cannot be measured by approaches used to test grown-ups. But in the past few decades, scientists have figured out better ways of getting information from infants. With these new tools, researchers have shown that babies' minds are very complex right out of the box—as many parents suspected all along.

All brains, young and old, have certain broad talents that help their owners to navigate life successfully. If you look closely, you can already see many of these talents in your infant. Although babies lack knowledge, they are born with certain tendencies that influence how they organize incoming information and respond to it. They are predisposed to seek out experiences that will help adapt their growing brains to their particular environment. Or, to put it more simply, your child's brain naturally knows how to get what it needs from the world. For this reason, most brain development requires only a "good enough" environment (more on that later), which includes a reasonably competent (though not perfect) caretaker.

What do babies know and when do they know it? They can't tell us in words, but researchers can still ask babies questions and get sophisticated answers about their **cognitive** abilities. A few simple, nonverbal ways of looking into the minds

of infants and even newborns have revolutionized developmental psychologists' ability to tell what young babies think and feel.

Your infant isn't good at controlling most of her body, but she can suck on a nipple immediately at birth. Not long after that, she can turn her head and eyes to look at an interesting object or event. These two abilities can be used to find out what catches her attention. For example, if your infant likes an event that happened while she was sucking and wants it to happen again, she will suck more vigorously. Your newborn will suck harder when she hears a recording of her mother speaking, but less so when she hears another woman. This is how we know that, from birth, infants recognize Mom's voice.

Like adults, babies get bored. After your baby has looked at something for a while, he will turn away and look at something more interesting. Researchers can observe how long a baby looks at a particular scene. If the scene contains something surprising to the baby, he will look longer.

This response allows us to find out whether a baby can tell the difference between two things. For example, if you show your baby a series of pictures of cats, the appearance of a dog will attract a long look. This means that babies can distinguish cats from dogs—a feat that is extremely difficult to program into a computer.

Simple tools like these enabled researchers to identify five brain talents that infants already have well before their first birthday.

The first talent: babies can detect how common or rare particular events are. For example, a first step in learning a language is figuring out which syllables go together to form a word. Yet when speaking, people tend not to pause between words. One way to learn words is to determine which syllables are likely to occur together. For example, when your baby hears the words *the baby* being spoken, how can she tell that it's the English word *the* followed by *baby*, and not the made-up word *theba* and then *by*? One clue is that *baby* is a far more common pairing of sounds than *theba*.

A well-designed experiment showed that in general, babies really do think this way. Researchers generated four nonsense words, such as *bidaku*, each composed of three syllables. They then presented these nonsense words to eight-month-old babies in varying order, without pauses between the words. Once the babies were familiar with these new words, the researchers then presented either one of the nonsense words or a new one composed from the original syllables (like *kudabi*). They let the babies control how long the words were played by looking in the direction of the speaker. The researchers found that babies listened significantly longer to the new words, even though the component syllables were the same. Since the babies had already heard all the syllables individually, the researchers concluded that they must have become familiar with the original groupings. This ability to detect the probability of events, shared by many animals, is a key component of learning. It provides the basis for answering important questions like "Where am I most likely to find food right now?"

The second talent: babies use coincidences to draw conclusions about cause and effect. After language develops, two-and-a-half-year-old children can make explicit causal statements like "He went to the refrigerator because he was hungry." But well before this, babies appear to be able to detect such relationships.

In one experiment, a mobile was hung over the crib of three-month-old babies and attached to one leg by a ribbon. When a baby kicked, the mobile would move. The babies were fascinated by this new toy. They smiled more and looked at the mobile more than they did when a similar mobile was out of their control. After just a few minutes of training, they kicked more. Three days later, they still kicked when they saw the first mobile (but not a different one), even when the ribbon was no longer tied to their legs. Since the kicking was a specific response intended to get the mobile to move, these babies seem to be learning an elementary form of cause and effect. Using events that occur together to determine possible underlying causes is a key part of our ability to learn how the world works.

Babies can distinguish cats from dogs—a feat that is extremely difficult to program into a computer.

The third talent: babies distinguish *objects* from *agents* and treat them very differently. Infants—like all other people—understand that objects are cohesive (all the parts of the object stick together), solid (something else can't go through an object), and continuous (all the parts of an object are connected to other parts), and only move when something touches them. For many years, it was accepted that infants under eighteen months did not understand *object permanence*, the idea that objects continue to exist even when you can't see them. This bit of popular wisdom, originally disseminated by the pioneering psychologist Jean Piaget, was recently challenged by researchers who found the right ways to test infants.

Well before their first birthday, infants look longer if an object fails to be cohesive, solid, continuous, or permanent. In one experiment, five-month-old babies saw a car roll down a track whose middle section was hidden behind a screen. When a boxlike obstruction was then placed on the track behind the screen, five-month-old babies appeared to expect it to stop the car. How do we know this? When researchers secretly removed the obstruction through a trapdoor, and the car continued to roll down the track successfully, the babies looked longer at the screen, suggesting that they were surprised that the box was not solid. When evaluated in this way, babies as young as three and a half months old show that they can think about objects that are out of view behind other objects.

MYTH: IF ANYTHING GOES WRONG, MOM IS TO BLAME

Sigmund Freud has a lot to answer for. His ideas were speculative and eventually discredited by further research, but they have left deep impressions on our culture. One of the most pervasive ideas is that a baby's relationship with his or her mother serves as a model for all later relationships in life. This idea has led many people to conclude that a mother's behavior has an incredibly strong influence on what kind of a person her child will later become. From this belief, a culture has arisen in which complete strangers feel a moral obligation to intervene if they see a pregnant woman having a sip of wine or a mother yelling at her young son. In the past, psychiatrists even blamed mothers for their children's autism or schizophrenia—both developmental disorders that are largely due to genetic mutations.

It's time to relax. Now that you know that children actively participate in their own development, it should be clear that parents do not need to be perfect. We don't recommend yelling, but that's mostly because it's an ineffective way to modify your child's behavior (see chapter 29), not because your occasional bad mood is likely to do any serious, lasting damage to his psyche. Anyway, as you'll see in chapter 17, parenting style has much less influence on personality than most of us believe. We'd like to see parents enjoying their kids more, rather than worrying over every aspect of their growth. That approach would be just as effective in producing healthy adults—and much more fun for everyone.

Babies also recognize agents, beings that have intentions and goals and can move on their own. Hands, for instance, always belong to agents. If six-month-old babies see a hand reaching for one of two objects, they seem to understand that the person wants that particular object. When the location of the objects is then reversed, the babies look longer if the hand reaches for the same location (but a different object) on the second try. If instead a stick pokes the object, babies don't act surprised when the stick fails to follow the object to a new location, because a stick is not expected to act like a conscious agent.

Like adults, babies are willing to attribute agency to things that are not really alive. When watching a film of a circle that seems to be chasing another circle, one-year-old babies look longer if the first circle moves away from the second circle than they do if the first circle moves straight toward its presumed target.

The fourth talent: babies organize information into categories and people into groups. When infants as young as three months see a series of male faces, they spend less time looking at each new face, presumably because they're bored with looking at men. When a female face then appears, they look longer. This is true even if the hair is not visible, so the babies seem to be using facial features, not hairstyles, to distinguish men from women. These categories are relevant for babies' everyday lives. Most babies prefer looking at female faces to looking at male faces—except when their primary caretaker is male, in which case they are able to muster a slight preference for men.

Some broad categories like *animals* and *furniture* can be found very early in life; others are learned later. The boundaries of many categories, from the sounds of language to face perception, are shaped by experience to match your child's local environment. But no one ever has to teach babies that categorizing things is a good strategy; it's built into their brains. This ability provides a primitive basis for adult categorization, which makes it possible to think sensibly about newly encountered objects and people. It is also the root of stereotyping and prejudice, as we will see in chapter 20.

The fifth talent: babies select relevant information for attention while discarding most of what goes on around them. As you may have noticed, babies are much less selective than adults about what captures their attention, but they still have distinct, automatic biases. From an early age, babies focus a lot on human voices, faces, and moving things. Babies start showing this preference for faces at thirty minutes after birth, and for human voices two days later. After three months of age, they notice objects that look distinctly different from surrounding objects, such as a red circle in a field of black circles.

Very early on, caregivers begin to influence the direction of a baby's attention. Babies start to follow an adult's gaze as early as four months of age. By twelve months they can point and direct their attention where someone else is pointing. At all ages, paying attention greatly increases the brain's ability to learn about specific things. In computer models of brain function, innate biases in what information is given priority can provide a powerful mechanism for directing the learning of particular tasks. Babies' innate interest in voices, for example, helps them to learn about language. All of these talents help babies' brains develop like dandelions, requiring only everyday types of stimulation that adults give normally—and instinctively.

In adults, these five talents are fundamental to the way our brains work. Indeed, in most of us these talents are inclined to be hyperactive. When we find ourselves considering our computers or our cars as if they had their own intentions and goals (typically in opposition to whatever we'd like them to do), our tendency to perceive agents is getting out of hand. When a baseball pitcher wins three games while wearing a certain pair of socks, and then insists on wearing his lucky socks whenever he plays, he is drawing conclusions about cause and effect from events that probably occurred together by chance.

There's a practical reason why many of our scientific examples come from three-month-olds: younger babies are harder to test. Based on the evidence we have, our own belief is that these capacities are present from birth, at least in some primitive form. In the end, though, we don't think it matters very much whether babies are born with these abilities or learn them soon after birth. Either way, babies start relying on these tools in infancy and continue to use them throughout their lives. On the other hand, these cognitive capacities are just the beginning. All of them become significantly more elaborate as babies grow and mature.

From an early age, babies focus a lot on human voices, faces, and moving things.

This emerging picture leaves little room for the outdated idea that babies are born with the potential to develop in any direction. Instead, they all start with certain biases. The cognitive talents that babies have in early life are essential for the development of their brains. Computer scientists who construct simulations to model what the brain does also confirm that biases are necessary to make these programs act realistically, even though our biases may limit us in some ways. They have not been able to explain convincingly how an adult brain might develop from a learning machine that starts with no predispositions.

As a consequence of these core talents, children's brains come ready to learn how to adapt themselves to the environment that they encounter during development. This ability allows children to grow almost anywhere. Our species has survived under a wide range of conditions through its history, and we have evolved to learn about the properties of the environment that were directly relevant to our survival. For this purpose, targeted learning mechanisms are often better than general mechanisms. These predispositions prepare the infant brain to learn many things—but not just anything.

Chapter 2

IN THE BEGINNING: PRENATAL DEVELOPMENT

AGES: CONCEPTION TO BIRTH

When we watch a house being constructed, we always find it surprising how quickly the framing is done. From the outside, the house looks almost complete very soon after it's begun, but finishing the interior details and the wiring takes much longer. Building a brain is similar: getting the signaling cells, called **neurons**, into their correct positions is the (relatively) easy part, and it's done before your baby is born. In contrast, wiring up all the connections is so complicated that the job won't be entirely finished until your child is in college.

A baby's brain is different from a house in one amazing way: from fertilized egg to newborn, its construction is largely automatic. The processes that form the brain are driven by a resilient genetic program, allowing babies to grow in almost any environment. Its main requirement is a healthy mother. As it says on the packaging of some of our favorite appliances, no assembly required.

This chapter will lay out some of the most valuable advice we have about prenatal development, including warnings about some of its hazards. But before we get into the details of what we found in the scientific literature, we want to emphasize this point: **most pregnancies turn out fine.** Authors of many popular advice books (you know who you are) convey the message that women must avoid any risky behavior while pregnant, no matter how minor. While long lists of potential problems may frighten mothers-to-be and help to sell books, those lists can also lead to prenatal stress, which itself is bad for the baby's development (see

Practical tip: Less stress, fewer problems, p. 16).

The effects of prenatal risks depend on their timing and seriousness. In most cases, when miscarriages or birth defects do occur, they are not caused by the pregnant woman's actions. Throughout your life, it will be tempting to blame yourself for anything that happens to your baby or child, but keeping a clear perspective is essential.

Knowing a bit more can help you relax a little about the largely self-organizing process of prenatal brain development. At the same time, there are a few simple points to remember in which your involvement could make a difference. It is a technical subject, but bear with us; the basics are not as complicated as they might first sound.

The construction of the brain begins early in pregnancy. During the first

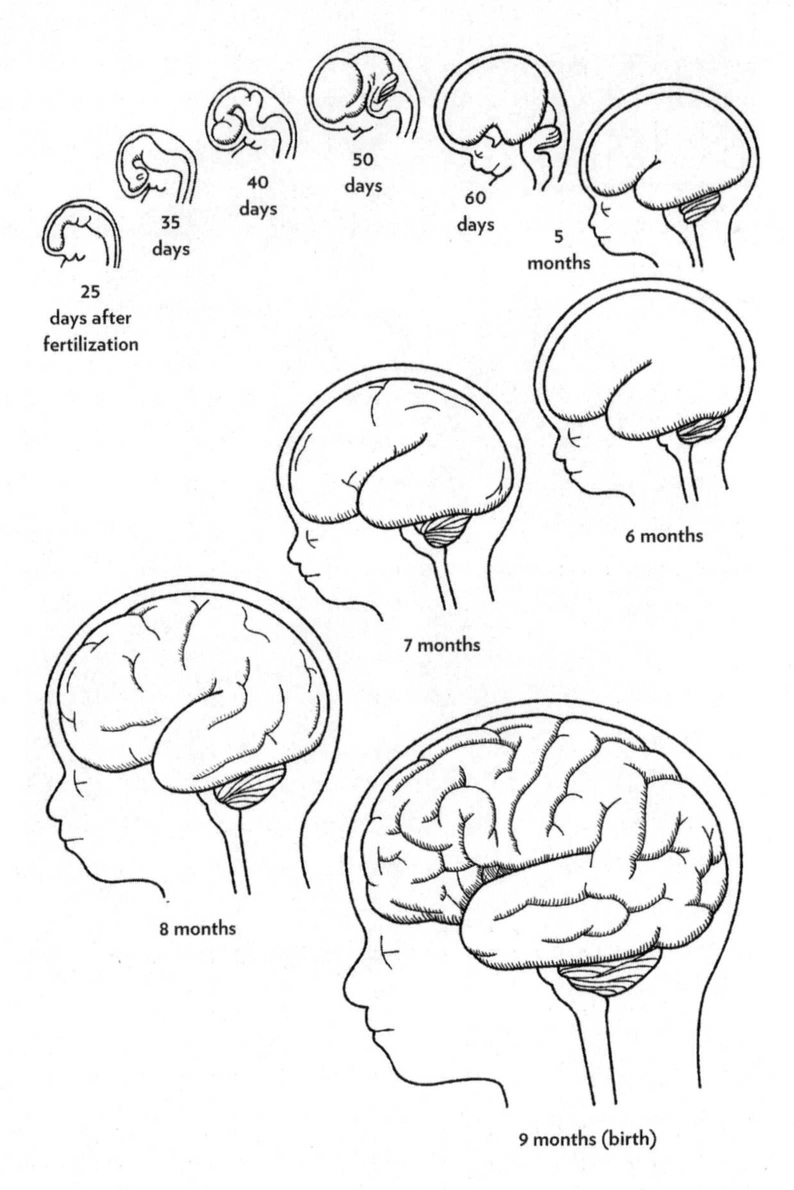

month, chemical signals cause a group of cells in the developing embryo to start becoming the nervous system. Beginning three weeks after conception, the *neural plate*, a cell layer running along the length of the embryo, brings its edges together to form the *neural tube*, which will later become the brain and spinal cord.

If the neural tube fails to close fully, miscarriage or birth defects such as spina bifida can result. (In spina bifida, the incomplete closure of the neural tube sometimes leads to the protrusion of the spinal cord from the vertebral column.) Folic acid deficiency in the mother increases the risk of such neural tube defects.

For this reason, women who might become pregnant should take 400 micrograms of folic acid (a B-complex vitamin) every day—or more if they're having more than one baby. Another source is bread, which in the United States and many other countries is made from folic acid–supplemented flour for this reason. If you are trying to have a baby, you should begin taking this supplement before conception, as many women do not find out they are pregnant until neural tube closure has ended, four weeks after conception.

The next stage of development is segmentation, which divides the neural tube into distinct regions by the sixth week of pregnancy. You can think of it as placing walls to define the rooms of a new house, except that segmentation is controlled not by physical barriers but by chemical cues. The largest neural tube region, at the back end of the human embryo, will become the spinal cord. A smaller area at the head end is divided into three regions, which will eventually become different parts of the brain (see figure opposite).

The hindmost of these three regions will become the **brainstem**, which controls mostly subconscious basic functions, such as reflexive movements of the head and eyes, breathing, heart rate, sleep, arousal, and digestion. It also forms the **cerebellum**, which integrates sensory information to help guide movement (for instance, so that you know how forcefully you need to lift your foot when walking).

The middle region will become the brain's midline structures, including the **hypothalamus**, **amygdala**, and **hippocampus** (see figure, p. 14). The hypothalamus controls many basic processes, such as the regulation of sexual behavior, hunger, thirst, body temperature, and daily sleep/wake rhythms, and the release of stress and sex hormones. Emotions, especially fear, are the responsibility of the amygdala. The hippocampus has two main functions: it stores information into long-term memory, and it is important for spatial navigation.

The third region, at the front of the brain, will become the **thalamus** and cerebral cortex, also called the **neocortex**. Sensory information entering the body through the eyes, ears, or skin travels to the thalamus, in the center of the brain, which filters the information and passes it along to the cortex. Scientists divide the cortex into four parts, or *lobes*. The **occipital lobe** is responsible for visual perception. The **temporal lobe** is involved in hearing, including the understanding of language. This lobe also interacts closely with the amygdala and hippocampus and is important for learning, memory, and emotional responses. The **parietal lobe** receives information from the skin senses, puts together information from

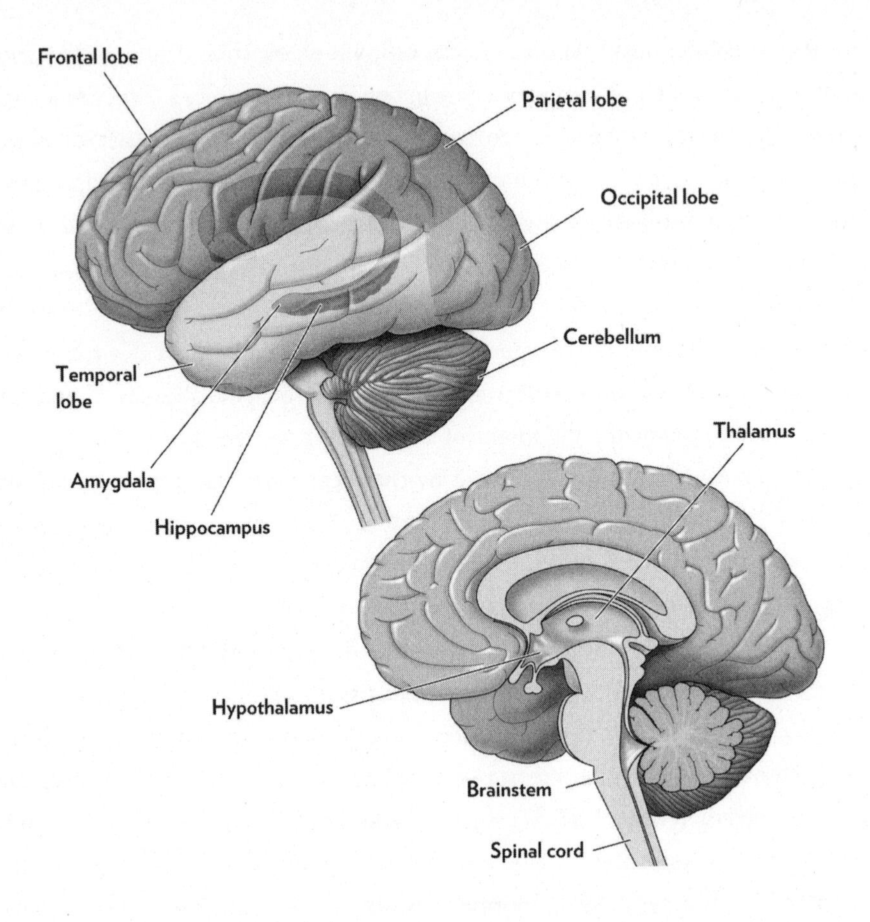

all the senses, and directs your attention. The **frontal lobe** generates movement commands, directs the production of speech, and is responsible for selecting appropriate behavior depending on your goals and your environment.

Early in gestation, all these brain regions are tiny. As development continues, chemical markers divide the brain into progressively more regions, defining particular cortical areas, such as those for certain aspects of vision or language. A cluster of cells with a common function is often called a **nucleus**. Once all the brain areas are specified, they grow larger, maturing in sequence from the back of the brain to the front (see figure, p. 12). This process continues through childhood and into adolescence (see chapter 9).

The main construction technique in the early stages of brain development is the production of new cells—billions and billions of them. Cells of the early nervous system divide repeatedly to make additional progenitor cells. These cells can even divide as they move, leaving trails of neurons behind them. Cell division also produces various types of **glial cell**, which contribute to brain function in many ways. One type of glial cell helps guide the placement of neurons early in development by extending long fibers that act as trails for neurons to follow.

The number of cell divisions and what type of cells they produce are tightly regulated by a combination of chemical signals, which vary across brain regions, and interactions with preexisting cells. The addition of new neurons is largely complete by about twenty weeks of **gestational age** (which is counted from the first day of the last menstrual period), or eighteen weeks after conception. A very small number of neurons continue to be generated even into adulthood, and new glial cells are generated throughout life.

During this period, cells are also beginning to differentiate, taking on particular jobs in the brain. Cells differentiate in a series of steps, as their jobs are slowly made more specific by increasingly restrictive chemical signals.

At a basic level, neurons have a lot in common (see figure, p. 19). They receive chemical signals called **neurotransmitters** that are released from other neurons. When neurotransmitter molecules bind to **receptors** on the **dendrites** of the neuron, electrical and chemical signals are generated that can spread—all the way to the cell body in the case of electrical signals. If enough electrical signals occur at the same time, the cell body can make an electrical impulse that is used to talk back to other neurons.

This output signal, called an **action potential** or spike, is conveyed down the **axon**, a very long, thin fiber that reaches from the brain to the target, as far away as the toe in some cases. Each neuron has a single axon, which often branches to reach multiple targets. Neurotransmitter molecules are contained in specialized areas at the ends of the axonal branches and released by the arrival of a spike. When a neurotransmitter binds to receptors on another neuron's dendrites, that target neuron may be electrically excited or inhibited, depending on the identity of the neurotransmitter. The point of connection between axon and dendrite is called a **synapse**. Final stages of differentiation often depend on neurons' interactions at synapses.

Glia also come in different flavors. Some glia wrap themselves around axons like the insulating plastic sheath on electrical wire, forming a layer called **myelin** to increase the speed of neural communication. Other glia line blood vessels to control which chemical signals are permitted to pass into and out of the brain. Still others form the brain's defense system, engulfing and removing foreign matter and debris from dying cells. Glia too become differentiated by exposure to chemical signals, generally a bit later than the neurons in the same areas.

The first step in the wiring process occurs before birth, as these billions of neurons extend axons toward their targets. Fortunately, distances are much shorter in the fetus than they would be in an adult. It also helps that brain tissue is less crowded than it will eventually become, just as it's easier to run electrical wires and plumbing in a house before the interior walls have been put up. Only the earliest-arriving axons must find their way by themselves, navigating via chemical signals or by finding particular guidepost cells.

Later axons extend along the pathways laid down by these early pioneer axons, just as you might guide a new wire through a bundle of previously installed wires, except that the new axon is actually being created as it progresses. A bundle of axons in the brain is called a **nerve**. A region at the tip of the elongating axon called the *growth cone* samples the environment within the brain in different directions by extending and retracting small protrusions, making it look as though the growth cone is sniffing out the correct path. Depending on their identity, these chemicals may either attract or repel the growth cone. Some can even cause it to abruptly change its responsiveness to other molecular cues, a form of sophisticated navigational logic.

Once an axon has found its approximate destination in the brain, it must pinpoint its target cells from among millions of candidates. This process starts with molecular cues that tell the axon to slow down and start exploring an area whose boundaries may be marked with a repellent signal to prevent the axon

PRACTICAL TIP: LESS STRESS, FEWER PROBLEMS

Next time you're stressing about your future child, ask yourself whether this stress is really necessary. Neuroscientists are able to discover what stress does by studying its effects on laboratory animals. Maternal stress increases the risk of a variety of problems, including cleft palate, depression-like behavior, a touchy stress-response system in adulthood (see chapter 26), and attention deficits and distractibility (see chapter 28). Stress hormones released by the mother animal act on the fetus directly and also reduce the placenta's ability to protect the fetus from these hormones in the future.

Because it would be highly unethical to stress pregnant women deliberately, most research in people has relied on looking for correlations, which is less reliable than experimental results (see *Did you know? Epidemiology is hard to interpret*, p. 262). Some recent studies have examined children born after their mothers experienced natural disasters during pregnancy. This type of study comes as close as is ethically possible to randomly placing women into stressed and unstressed groups.

One group of researchers identified all tropical storms or hurricanes that hit Louisiana between 1980 and 1995 and then determined how many autistic children in the records of the state health system had been in the womb when their mother's home was hit by one of these storms. The risk of autism was significantly higher for children whose mothers had been stressed during pregnancy—though most cases of autism probably result from other causes (see chapter 27).

By scientific standards, this evidence is far from ironclad, but there are two reasons to believe it's not mere chance. First, the incidence of autism was higher only for those children whose mothers were in the fifth, sixth, or ninth month of pregnancy at the time of the hurricane, suggesting that there is a period when the effects of stress on development are long-lasting (see chapter 5). Second, children whose mothers were exposed to more severe storms had a higher risk of autism than children whose mothers were exposed to less severe storms. This research will need to be replicated before we can consider it definitive, but it does suggest that prenatal stress may increase the chances of autism.

Similar studies have yielded comparable results. One found that children whose mothers experienced severe stress from a major ice storm while pregnant had lower intelligence quotient (IQ) scores and language ability at age five. The risk of schizophrenia is higher in children whose mothers were in the first trimester of pregnancy when a close relative died or was diagnosed with a serious illness. Children whose mothers experienced an earthquake during pregnancy were more likely to be diagnosed with depression or to be born with a cleft palate. It's not yet clear whether moderate stress, such as dealing with an annoying boss, might cause similar problems, but as the research is ongoing, it's best to keep it simple: it's probably a good idea to take time to relax and be kind to yourself during pregnancy as much as you can.

from exiting. Some brain areas help the axon to navigate by providing a local map, in which the concentration of a chemical signal (or several) descends steadily across the area. Other areas use a large number of related **proteins** to mark local position so axons can find their way to the right neurons. Proteins are the universal building blocks made by cells for a wide variety of functions. In this case, the function is to say to an axon, "You are here."

Once axons are close to their destination, they begin to make contacts with nearby cells, initiating the chemical conversation that leads to the formation of synapses. This process begins in the spinal cord five weeks after fertilization, and it is not complete until years after birth in some brain areas. Axons initially form a lot of extra synapses with targets that are only roughly appropriate. Only some of these synapses survive in the long term. Synapses that are more successful at activating their target cells are more likely to be retained. This competition for synaptic survival provides a way of fine-tuning the brain's function to match each child's individual circumstances, whether that means adapting the responses of vision neurons to the distance between each child's eyes or tuning the auditory cortex to respond most easily to the sounds of each child's native language. To a lesser extent, this process will continue throughout life, as a mechanism of learning and memory (see chapter 21).

The process of eliminating unnecessary components is a major theme of early development. The adult brain contains about 100 billion neurons and many more glia. However, the young brain produces even more cells than that and then reduces their number through planned cell death. In some brain regions, planned death kills as many as four out of every five cells born. These events are called *regressive* by neuroscientists, and they are essential for normal development.

Why does the nervous system take such a wasteful approach? It seems to be a way of matching the size of the incoming population of axons to the number of neurons in the target region. Cell death occurs after the axons have reached their target and formed synapses. The target neurons produce a protein, necessary for cell survival, which is taken up at synapses and transported back along the axon to the cell body of the input neuron. Cells that have failed to form enough connections with the target do not get enough of the survival substance, so they die. This type of cell death is an active process, resulting from a biochemical death pathway within the cell. The best-known survival protein (or **neurotrophin**) is nerve growth factor, which controls the survival of neurons involved in the sense of touch and

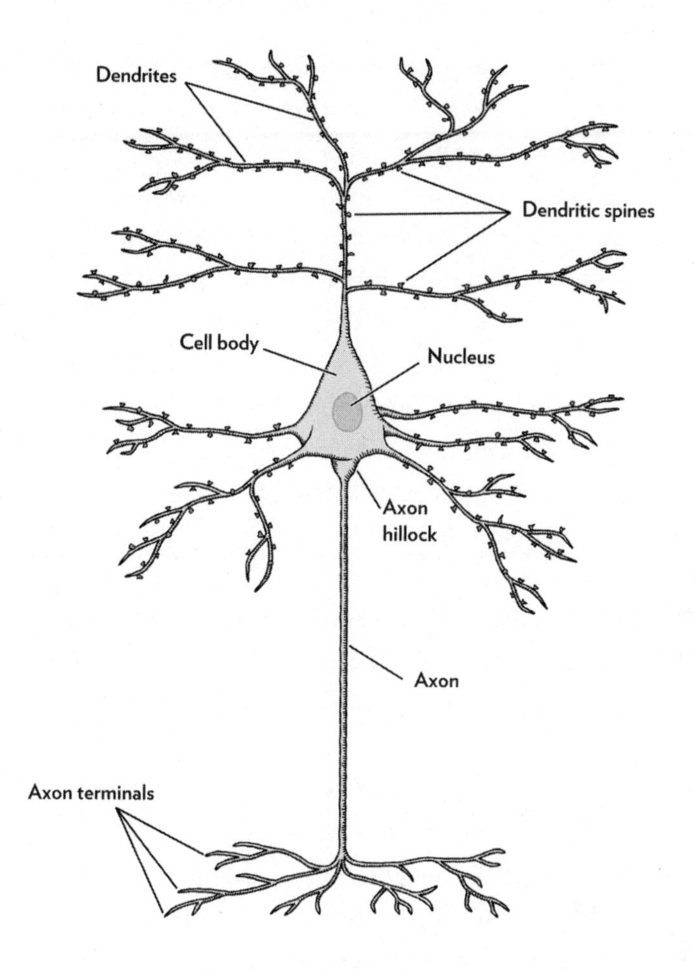

the fight-or-flight reflex in the peripheral nervous system. Other factors also influence cell survival, including incoming synaptic activity and sex hormones, which control cell death in brain regions that differ between males and females.

Even after all the cellular elements of the brain are in place, much construction work remains to be done. Newborn neurons look very simple compared with mature neurons. Toward the end of gestation and especially in the first two years of life, dendrites form additional branches, becoming more and more complex to accommodate the many new synapses that are added during this period. Synapse elimination begins in the first year of life and continues through early adolescence, forming one of the basic mechanisms by which experience helps to shape the brain (see chapter 5).

The final step in axon maturation is myelination, the formation of the glial insulation that allows spikes to move quickly down the axon. It's as if the brain's electrical system were installed with bare wires, and then the insulation got added afterward. This process begins just before birth in the brain (earlier in the spinal cord) and continues well into adult life (see chapter 9).

Considering the enormous amount of construction involved, it's no surprise that the growing baby requires energy. Indeed, one of the biggest threats to a developing fetus is maternal malnutrition, whether caused by famine, poverty, or dieting. A particularly critical time is the second and third trimesters, when brain size is increasing rapidly. An unexpectedly low birth weight (compared with the baby's expected genetic growth potential) signals a higher risk of many problems later in life, including deficits in cognitive development and intelligence. Low birth weight and other problems, such as mental retardation and inflammation of the **retina**, are also associated with a variety of viral infections, including toxoplasmosis, rubella, and herpes simplex. In general, it is prudent to practice good hygiene late in pregnancy, and for that matter around young babies. High birth weight can be unhealthy as well. Customized growth curves do exist. Ask your obstetrician about them.

During pregnancy, environmental toxins can be a threat if they are ingested. For instance, cocaine use increases the risk of attention-deficit/hyperactivity disorder (ADHD). However, more severe effects on the brain result from two legal drugs, nicotine and alcohol. Low birth weight and a variety of brain development problems are linked to smoking, the nicotine patch, and heavy drinking. So-called crack babies, whose plight got a lot of press in the 1980s, turned out to be damaged mainly by their mothers' malnutrition and concurrent use of other drugs.

In the *Mad Men* era, the sight of a pregnant woman with a drink in one hand and a cigarette in the other did not attract a second glance. Today, some U.S. states jail women for child abuse if they are caught taking cocaine while pregnant, but not for smoking or alcoholism. To put it mildly, this approach is not optimal for the baby's health (see *Practical tip: Less stress, fewer problems*, p. 16). Although

PRACTICAL TIP: EAT FISH DURING PREGNANCY

Exposure to lead (from paint, water pipes, dishes, and even imported makeup) or mercury in utero or in childhood can decrease intelligence. These heavy metals are harmful to brain development. For years, women were told to limit their con-

sumption of fish because it might contain mercury. But fish is also a major source of omega-3 fatty acids, which are crucial for neural development. Indeed, their absence during brain formation can lead to mental retardation. To settle this matter, we dug into the scientific literature to weigh the risks and benefits. Our verdict: fish is good. Several long-term studies now show that children whose mothers eat fish during pregnancy have better-functioning brains than children whose mothers avoid fish—especially if the mother chooses fish species that are low in mercury.

One group of researchers evaluated the eating habits of 11,875 women living in Bristol, United Kingdom, during the third trimester of pregnancy and then tested the resulting children on a variety of cognitive measures. Mothers who avoided seafood were more likely to have children with poor social behavior (at age seven) and low verbal IQ (at age eight) than mothers who ate at least three six-ounce (170-gram) servings per week. The more seafood a mother ate, the better her child's brain functioned, which suggests that the effect was due to the fish itself and not to related characteristics such as household wealth. The benefit is small but well documented. (A statistical estimate of effect size is about 0.2 to 0.3; see chapter 8 for a discussion of this measure.) No benefits were seen among mothers who ate fewer than two servings a week. Another study confirmed these findings and further showed that children of mothers who ate fish low in mercury during pregnancy had higher verbal intelligence than children whose mothers ate fish high in mercury, for the same fish consumption.

You may have heard that uncooked fish, especially wild Pacific salmon, can contain disease-causing parasites. Thorough freezing of the fish to kill the parasites, which is required of sushi consumed in the United States, minimizes this risk.

How do you know if your fish contains mercury? A good rule of thumb is the smaller the fish, the less mercury it is likely to contain. Top predators like swordfish and shark should be avoided because mercury and other contaminants get concentrated as they go up the food chain. Your health department may have information on the risks associated with your local fish. The most important point, though, is that the benefit of getting enough omega-3 fatty acids seems to outweigh the risk of mercury contamination for fetal brain development.

prison time for lighting up while pregnant is not likely anytime soon, you can improve the health of your baby by addressing these habits early in pregnancy—or better yet beforehand.

Another source of drug exposure during pregnancy is medical care. Pregnant mothers are advised to avoid a variety of over-the-counter drugs. In regard to brain growth, the third trimester is a time of vulnerability. Drugs can enter the placenta, and therefore the baby's developing brain, and increase the likelihood of neurodevelopmental problems. One drug given at this stage is terbutaline, an activator of receptors for the neurotransmitter **epinephrine**, which is intended to prevent preterm labor—but is in fact ineffective at doing so. Steroids that emulate the stress response are given to improve lung development in babies at risk for preterm birth, but multiple courses of steroids can harm the developing brain. Even for drugs that are currently regarded as safe, the possibility of a risk exists late in pregnancy, when the baby's brain is growing rapidly. Though in many cases there is no alternative, risks and demonstrated benefits for both mother and baby should be weighed carefully.

The news is not all bad for drugs, though. One of our favorites, caffeine, is harmless in moderate doses of no more than 300 milligrams a day (three regular cups of coffee—or a single Starbucks grande), as are artificial sweeteners and monosodium glutamate (better known as MSG). Although it is generally not recommended, there are some doctors who even approve of alcohol in small doses. So expectant mothers don't need to give up all of their favorite habits. Indeed, a little less worry on this front might be a stress reducer.

An important threat to the baby occurs when the pregnancy cannot run its full course. A common cause of low birth weight is premature birth, which greatly increases the risk of neurodevelopmental disorders. One Norwegian study found that babies born during gestational weeks twenty-eight to thirty have a fourfold higher incidence of mental retardation, a sevenfold higher incidence of autism spectrum disorder (see chapter 27), and a forty-six-fold higher rate of cerebral palsy than full-term babies. By the age of eighteen, one in twelve of these children was classified as disabled—five times the normal rate. Premature babies born later in gestation have lower rates of disability, but even at thirty-seven weeks, the risks remain elevated over babies born at full term. A major contributor to preterm birth is carrying more than one baby. Couples using in vitro fertilization can reduce this risk by asking their doctors to transfer no more than one embryo per cycle.

One unintended consequence of recent changes in medical practice is an increase in the frequency of premature birth. Preterm babies (less than thirty-seven weeks) now make up 12–13 percent of births in the U.S. This percentage rose steadily between 1981 and 2004, partly because their survival rates grew as medical care improved. Three quarters of preterm births occur late in gestation, between thirty-four and thirty-seven weeks. Among babies born in this time range, 20 percent end up with clinically significant behavior problems, and the risk of ADHD is 80 percent higher than in full-term babies.

A recent study suggests that as many as one in fifteen induced deliveries between thirty-four and thirty-seven weeks are done without a compelling medical reason. Examples include hypertension or a complication in a prior pregnancy, neither of which is considered an absolute indication for early induction. In these cases, early delivery should be weighed against the risks we have described. Of course, in many cases, such as bleeding or a prolapsed umbilical cord, the necessity is unavoidable.

One of the biggest threats to a developing fetus is maternal malnutrition, whether caused by famine, poverty, or dieting.

Therefore one of the best things you can do to protect your baby's growing brain is to allow your pregnancy to run its full course whenever possible. Between 10 and 20 percent of all U.S. deliveries are induced, and roughly half of these inductions are elective (not medically necessary). Reasons for elective induction include doctor or patient convenience and doctors' concerns about legal liability. We recommend following the American College of Obstetricians and Gynecologists recommendation that elective induction should not be scheduled before week thirty-nine of pregnancy (this cutoff is set late because the exact stage of pregnancy is not always known with precision) unless there is a clear medical need. To use a baking metaphor, that bun in the oven will turn out great—if you wait until it's done.

Chapter 3

BABY, YOU WERE BORN TO LEARN

AGES: BIRTH TO TWO YEARS

No wonder babies sleep so much. They've got a lot of hard work ahead of them. Infants come equipped with a basic toolkit for learning, as we described in chapter 1. But that still leaves a lot of items on their to-do list. In the first year of life, babies must lay the foundations for all their adult abilities, from language to locomotion. Their brains are changing more quickly at this age than they ever will again. Many of those changes help babies to learn about the specific environment into which they have been born.

People can live in an astounding variety of places, from the frozen tundra to the sweltering desert, and in a vast array of social systems as well. Growing up in New York City or Barcelona is a very different experience from growing up in a subsistence village in the Amazon, but babies come into both of those situations with nearly all the same genes.

Unlike many animals, people are not hardwired to be a good fit to their environment at birth. Instead, babies arrive equipped with the skills required to adapt flexibly to a wide range of conditions, an ability that has allowed people to survive all over the world. The benefits of that approach are enormous, and so are the costs: children need a lot of care for a long time before they become independent. This high-risk, high-reward reproductive strategy affects the shape of most people's lives for decades, first as children and then as parents.

Babies are driven to explore and test their ideas about the world—which is why they seem to be getting into things all the time—and they love making things happen. When a baby learns to push a bowl from her high chair to make

a crashing mess, you can see the glee as she triumphantly proceeds to do it again and again. Being effective in the world is enormously rewarding for children and adults alike. Infants, though, sometimes get confused about how they caused something to happen, so you can see them trying to coax an object into behaving by talking to it. This confusion between physical and psychological causality usually disappears by the first birthday.

Just as babies have been shaped by evolution to be supereffective learners, adults have become equally effective teachers. It may look like a game of peekaboo, but there's serious stuff going on here. Babies are extremely good at getting what they need from their adult caregivers—not only food and shelter, but also

MYTH: **BREAST-FEEDING INCREASES INTELLIGENCE**

Everyone seems sure that giving breast milk to babies will make them smarter. We thought so ourselves when we started writing this book, but our careful examination of the scientific literature shows that the evidence for this idea is questionable.

There's no doubt that children who were fed exclusively on breast milk during infancy have higher intelligence on average than children who were not breast-fed. The important question is why this correlation occurs. One possibility is that this difference has something to do with the characteristics of breast-feeding mothers.

Indeed, women who choose to breast-feed their babies differ in many relevant ways from women who don't breast-feed. Compared to women who bottle-feed their children, women who breast-feed have higher IQs, are more educated, are less likely to be poor, and are less likely to smoke. An increase of about fifteen points (one **standard deviation**) in the mother's IQ more than doubles her likelihood of breast-feeding her baby. Headlines reading "Smart Mothers Found to Have Smart Babies" probably wouldn't be so memorable.

Researchers have tried to deal with these confounding factors in a variety of ways. **Meta-analysis**, a powerful statistical technique for combining the findings of multiple studies to increase our confidence in the conclusions, has produced inconsistent results. Some papers report a small effect of breast-feeding on IQ, and others find no effect. In general, though, studies with a large number of subjects have tended to find smaller effects. To a scientific reader, this is not encouraging news. Real effects (not due to chance) should be easier to see in large populations. One meta-analysis concluded that higher-quality studies were also less likely to find an effect of breast-feeding on intelligence.

In one large study, the IQ differences associated with breast-feeding were completely eliminated by taking the mothers' characteristics into account. Among the 332 pairs of siblings in which one was breast-fed and the other was bottle-fed, there was no difference in IQ. Another study of 288 sibling pairs in which only one child was breast-fed reported similar findings.

The ideal way to address these concerns would be to assign some infants to be breast-fed and others to be bottle-fed. One large study in Belarus came

as close as ethically possible by randomly assigning some mothers to a support program that increased the duration of successful breast-feeding. The authors reported that the support program substantially increased children's IQs at age six. Unfortunately, their IQs were tested by pediatricians who had a stake in the program's success and did not normally administer IQ tests. Indeed, when psychologists retested some of the children, their scores were significantly lower, raising the possibility of bias in the initial measurements and causing considerable uncertainty about the true size of the effects.

Our reading of the literature is that the weight of the evidence suggests that breast-feeding has little or no influence on a baby's later intelligence. Of course, we're not arguing against breast-feeding, which has many other benefits, including the opportunity for loving physical contact with your baby (see chapter 11). But mothers who are unable to breast-feed should not worry that they are harming their baby's intellectual development.

---

information and examples. As a mother coos to her baby that he's such a good boy, he is learning about language, relationships, and much more.

Because of the brain talents we discussed in chapter 1, even newborns are not passive recipients of adult instruction. Instead babies actively seek out the information that is most useful to them at a particular stage of development, and their behavior reliably elicits the kind of help that they need from adults. For instance, many people speak to babies in **motherese**—a high-pitched, sing-song, and slow version of regular language with elongated vowel sounds. Babies prefer to hear motherese and interact more intensely with people who speak this way, as most adults and older children do instinctively. It is probably not a coincidence that the properties of motherese, including clear pronunciation and pauses between words, are also very well suited for helping babies learn about language.

Of course these interactions with adults influence some aspects of how babies develop, such as determining which language the baby learns. All normal babies eventually learn the things they need to know, but the rates and details of learning depend on the experience of growing up in a particular culture. For example, there are a lot of cross-cultural variations in timing and even occurrence of the stages of motor development that pediatricians use to determine whether your child is progressing normally (see *Practical tip: Guided practice can accelerate motor*

PRACTICAL TIP: GUIDED PRACTICE CAN ACCELERATE MOTOR DEVELOPMENT

Infants learn to hold their heads up, sit, and walk months earlier in cultures that provide a lot of tactile stimulation and help babies to practice motor skills. In African, Caribbean, and Indian cultures, mothers massage and stretch infants after bathing them. These routines can include swinging infants around or tossing them in the air. Babies carried in a sling improve muscle strength and coordination as they adjust to Mom's movements. Laboratory studies verify that such stimulation promotes motor development. Spinning an infant in an office chair twenty times twice a week over four weeks (a safe and fun way to provide stimulation) or moving the legs passively (twenty minutes daily for eight weeks) speeds the infant's acquisition of motor skills.

To teach sitting, some African and Caribbean mothers hold young infants in the sitting position on their laps or prop them up with a cushion or other support. To promote walking, mothers hold infants in a standing position and bounce them up and down, causing them to make stepping movements. Once babies are able to hold themselves up, mothers encourage the babies to take steps while leaning against furniture, sometimes luring them with food. In such cultures, even young babies spend much of their time in sitting or standing positions. Western parents take such a deliberately planned approach only when teaching their toddlers how to climb stairs.

Trained infants develop motor skills more quickly than untrained infants, but only the skills that are specifically practiced. Laboratory studies verify that babies who practice crawling movements start to crawl earlier, and babies who practice stepping start to walk earlier. In contrast, babies are slower to develop motor skills if their movements are restricted. Denver babies born in winter start to crawl the following summer three weeks younger than summer-born babies who learn to crawl in the winter, apparently because cold weather limits the latter group's opportunities to practice.

Do babies who walk earlier end up with better motor skills than those who walk later? Probably not, unless they continue to practice these skills more than other children. In some cultures, adults routinely run long distances or carry huge weights, but those skills require many years of training.

development). In the U.S., learning to crawl is widely considered a prerequisite for learning to walk, but it is merely one of many ways to get around that infants may discover. Almost a third of babies in Jamaica do not crawl at all, and the rest begin to crawl at the same age as first walking, around ten months. Similarly, 17 percent of British infants never crawl, and a hundred years ago, 40 percent of U.S. infants did not learn to crawl, probably because babies in that era wore long gowns that made crawling difficult.

Similarly, experience with language influences the development of concepts. The Korean language includes a complex system of verb endings to carry information, whereas English relies heavily on nouns to convey meaning. Korean baby talk is full of verbs that contain implied prepositions (*moving into*) and often omits nouns entirely, while English baby talk contains a lot of nouns (*a doggy*). Perhaps because of this experience, American toddlers begin to categorize objects, for instance, by sorting them into piles by type, at a younger age than Korean children. In contrast, Korean children learn to use a rake to retrieve a toy that's out of reach earlier than they learn to categorize, suggesting that they find it easier to think about actions than about objects.

From birth, babies can imitate other people and seem to enjoy doing it, which not only is a powerful tool of social bonding but also provides direct examples of behavior for babies to copy. Infants imitate the goals of actions, rather than their exact form, and other people's movements appear to be coded in their brains in terms of goals as well. For example, if a fourteen-month-old baby watches a person tap her head on a box, which then lights up, a week later he will tap his head on the box to make it light up. But if the demonstrator's hands were wrapped up in a blanket when she used her head, most babies will instead touch the machine with their hands, apparently assuming that the demonstrator used her head because her hands weren't available. You might amuse yourself by cooking up a similar game to play with your own baby.

During this period of intense learning, a huge number of connections between neurons are added to the baby's brain. Just before and after birth, as many as forty thousand new synapses are added every second. A baby's brain reaches 70 percent of adult size by the first birthday and 80 percent of adult size by the second birthday. This growth is pronounced in the cerebellum, a region that integrates sensory information to help guide movement, as babies are learning how to control their bodies. The cerebral cortex also has a lot of growing left to do at birth. It doubles in size over the first two years of life, with most of that growth happening before age one. Though a small part of the growth is due to birth of new neurons, most of it is caused by the formation of new connections. The elaboration of axons, dendrites, spines, and synapses, all parts of a neuron that allow it to talk to other neurons, occurs rapidly throughout the first year. Myelination of axons is also intense during this time, as glial cells wrap themselves around axons to form an insulating layer that increases the speed and efficiency with which signals are carried from one neuron to another.

You might imagine that a baby's experiences would determine where new synapses are formed, but that doesn't seem to be what happens. Instead the brain produces a huge number of relatively nonselective connections between neurons in early development and then gradually removes the ones that aren't used often enough (see chapter 5). If the brain were a rosebush, life experience would be the pruning system, not the fertilizer.

Babies are extremely good at getting what they need from their adult caregivers.

Motor development occurs in a sequence that is determined by brain maturation. Because the primary motor cortex contains a map of the body that develops in sequence, babies learn to control their head and face movements before they learn to reach, and only later do they learn to walk. By the third month of life, the infant's brain has developed enough to produce significant advances in behavioral control. At this age, babies start to be able to inhibit reflexes and eye movements. Their motor abilities allow them to react to maintain equilibrium when their posture is disturbed. They also develop clearly goal-directed behaviors, including head–eye coordination and reaching for objects. This transition also reduces the amount of time that babies spend crying. Fortunately for the parents of fussy newborns, behavior in the first three months of life does not predict future temperament very well.

By four months of age, the eye movements of babies show that they can predict when an object will emerge from behind a screen, the earliest exercise of a skill that becomes increasingly important with age. Learning to anticipate future events, such as correcting your posture to offset a threat to balance before it occurs, is a key aspect of adult motor function. Predictive motor control is another function of the cerebellum, so its maturation is likely to be important for the development of this ability.

Even when they're very young, babies know something about objects, but they still have much more to learn. The fact that space is three-dimensional seems to be apparent even to young babies. Newborns will duck away from objects that are heading toward them, and as soon as they can control their arms, babies will try to reach in the direction of objects that they desire.

The idea that objects have fixed properties, on the other hand, seems to dawn slowly. In early life, motion seems to be a key to object perception. Adults use this cue too—things that move together are seen to be parts of the same object—but babies take the idea to the extreme. At five months, babies who are shown a stuffed animal going behind a screen and a toy car emerging on the same trajectory do not appear surprised. At that age they can certainly tell the difference between the two toys, but the object's motion appears to be more important to them. By the age of one, changes in most object properties (such as shape or color) will elicit a reaction, suggesting that the brain's representation of occluded objects has become much richer.

Babies do all this work without needing any special classes or equipment. Any baby with a normal brain and environment can develop the skills that are important during this period of life. They are driven to practice these skills, and parents are well suited to teach them, just by interacting with their children in everyday life. Most parents of infants can simply do what comes naturally and enjoy watching and helping their babies make discoveries about the world.

Chapter 4

BEYOND NATURE VERSUS NURTURE

AGES: CONCEPTION TO COLLEGE

When scientists say that the "nature versus nurture" controversy is outdated because both forces work together, that's not a case of fatigued combatants pleading, "Can't we all just get along?" It's a biological fact—and we understand quite a bit about how the process works.

So far, we have told you that your child's brain builds itself through largely automatic programs, and that it adapts to its environment. These two statements might seem to contradict each other if you think of *automatic programs* as *genes*, as some people may be tempted to do. That's not quite right, though; these programs control the interplay between genes and the environment during your child's development.

One reason that people get so worked up over this debate is the widespread assumption that genetic contributions to development are deterministic, while environmental contributions are flexible. That's why it's seen as conservative to argue that boys and girls are biologically distinct, and as liberal to argue that socialization is responsible for their behavioral differences. Such discussions lead nowhere because both assumptions are incorrect.

Genes establish a program to build a brain (see chapter 2), but then that brain reacts to the world, extensively tuning itself to local conditions as your child grows. The human ability to live in a wide variety of circumstances has resulted from natural selection favoring genes that contribute to behavioral flexibility (see *Did you know? Culture can drive evolution*, p. 36). Nearly all genes that influence behavior act by changing the odds of a particular developmental outcome, not by specifying it exactly—so your child's heredity is not destiny.

On the other hand, some environmental effects on development cannot be undone. For instance, the surrounding culture completely determines which native language your child will speak, but once the learning process is finished, there's no possibility of substituting a different native language in its place.

Indeed, from an individual neuron's perspective, it would be hard to distinguish between "genetic" and "environmental" influences. Signals that enter your brain through your eyes or ears (that is, via experience) influence development by causing chemical signals to modify genes or proteins—just as genetic influences do. Some of these changes are reversible, and some are not, but whether they originated inside or outside the body is not the determining factor.

Later in this book we will talk about how experience can change the connections and chemistry of neurons. Here we want to explain just how entangled genes and environment are in your child's development.

DID YOU KNOW? FOOTPRINTS ON THE GENOME

How can your child's experiences cause permanent changes to her genes? This idea may seem to fly in the face of what you learned in science class, but it relies on cellular processes that are familiar to biologists. In response to a variety of signals, so-called **epigenetic** modifications can silence a region of DNA by making chemical changes that affect its shape, so that the protein encoded by that gene cannot be made (see figure opposite). When DNA is copied during cell division, the pattern of epigenetic modification is copied as well, so that all descendants of the cell maintain the information.

Researchers have long known that this process explains why various cell types (such as neurons versus kidney cells) look and act very different, even though they contain the same DNA. More recent work reveals that environmental events can cause similar long-lasting changes to DNA, providing a way for transient experiences to permanently modify gene expression. The accumulation of epigenetic modifications also explains why identical twins, who share all their genes, do not look exactly the same.

When epigenetic modifications occur in sperm or eggs, they can affect future generations. This process is best understood in laboratory animals. For example, female mice that spent a particular two weeks of their youth in an "enriched environment" (with many toys) learned more easily as adults. And so did their pups—even when those pups were raised by a foster mother and did not receive any enrichment themselves. The pups instead benefited from their mother's experience, passed down through epigenetic modifications to her DNA.

This research is in an early stage, so the list of known epigenetic effects continues to grow. Early social experience can modify later behavior, including stress responsiveness (see chapter 26), due to epigenetic modification of particular genes. Prenatal and early postnatal nutrition can influence the adult risk of heart disease, type 2 diabetes, obesity, and cancer in people. Experiments in animals support the idea that these effects may be due to epigenetic modification of DNA. Cocaine addiction also seems to involve epigenetic changes, perhaps explaining why it is so difficult to reverse. Epigenetic modification is a simple chemical process, but it can encode life's experiences, quite literally, into who you are, and even who your children will become.

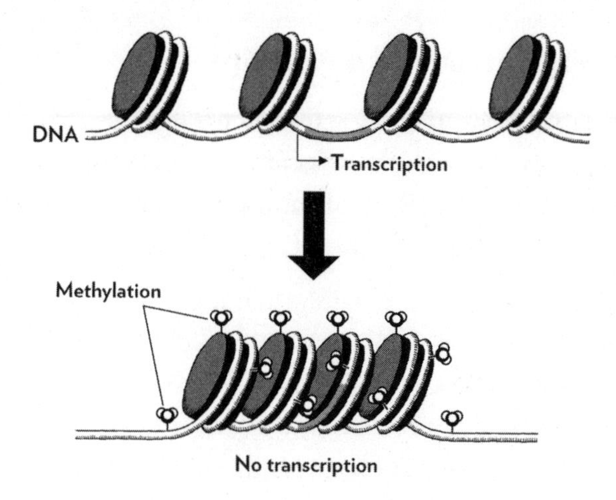

To start with a basic principle, your child's genes can influence his environment—and vice versa. His personal characteristics lead him to seek out certain experiences in life (see *Did you know? Why you're turning into your mother*, p. 150), and his tendencies to react to other people in certain ways affect how they behave toward him. A fussy infant who is difficult to soothe cannot be treated exactly like a calm and happy infant, no matter who the caretaker is. At the same time, the environment that your child encounters (before or after birth) can cause permanent changes to his genes. Chemical modifications in response to experience, such as methylation, can turn off certain genes in particular cells—often for a lifetime (see *Did you know? Footprints on the genome*).

Because the influences run in both directions, many developmental processes are feedback loops, in which our genes influence our environment, which then influences our genes (or at least the way they are expressed), and so on. The idea of reciprocal influence is challenging to grasp. When we think about genetic inheritance, our minds usually reach for a familiar example, such as our schoolhouse lessons on Gregor Mendel's wrinkled and smooth peas, or in our lives, genes for eye color. These examples are taught in school because they act in simple ways that can be drawn on a chalkboard, but the ways that most genes influence the brain and behavior are much more complicated.

There is no doubt that both genes and environment strongly influence individual differences in behavior. The genes that a baby inherits from her parents do not determine exactly what kind of person she will become, but they do define the range of possible developmental outcomes that are open to her. Even so, the same genetic tendencies can play out very differently in different cultures (see chapter 20). With all this interaction, it is nearly impossible to figure out how much of a particular behavior is caused by genes and how much by environment.

The first problem is that these kinds of estimates apply only to one particular environment that was studied and may change (a lot) under other circumstances.

DID YOU KNOW? CULTURE CAN DRIVE EVOLUTION

Our life experiences can also modify the human genome over evolutionary time scales—through the effects of cultural changes on natural selection. Geneticists think of culture as any learned information that influences behavior, which can include beliefs, values, skills, and knowledge.

When people first learned to domesticate cows around eight to nine thousand years ago in Egypt, milk became available as food for adults, not just infants. Before this change in the environment, which was entirely due to a cultural innovation, adults were lactose intolerant and could not digest milk. As herding spread, the genes that lead to production of the lactase enzyme in adults became more and more common because adults who could digest milk had access to a better food supply than those who couldn't. Different genetic mutations led to lactose tolerance in European and East African populations, in both cases around the time that herding was introduced. Today, lactose intolerance remains common in people of Asian and West African descent, whose ancestors did not adopt cattle herding.

Researchers estimate that between several hundred and two thousand human genes show signs of recent rapid evolution, many of which may have been driven by cultural changes. Genes that help our bodies respond to pathogens, for instance, have changed quickly. Evolution of new immune-system defenses may have been driven by the development of farming, which led to many new human diseases by bringing people into close contact with animals and their germs.

Other groups of genes are rapidly changing, too. One is related to digestion of various types of food and alcohol, which may be shaped by dietary practices. The invention of cooking was correlated with changes in digestion, bitter taste receptors, tooth enamel, and jaw musculature. Another example is genes that affect brain function, as we would expect from the substantial advantages provided by brain talents like language and learning ability. Somewhat curiously, a third category is genes that drive physical appearance, such as skin color, hair color and thickness, and eye color. Selection for these genes may be a product of culturally driven sexual preferences as well as environmental drivers such as sun exposure.

Cultural changes can also protect human populations from the selection pressures imposed by new environments. When people migrated to cold places, they learned to build fires and dress in fur coats, rather than developing the fur and insulating layers of fat that protect other animals from freezing temperatures. Some researchers have speculated that the ability to adapt to new environments through learned behaviors could free our species from some of the constraints imposed on other animals by natural selection, thus allowing us to maintain an unusually large variety of genetic traits in our population—another possible contributor to our famous behavioral flexibility.

For example, in middle-class populations, about 60 percent of the individual variability in IQ is attributed to genetic differences and almost none to the environmental circumstances shared by children within a family. In contrast, among people living in poverty (see chapter 30), about 60 percent of individual variability in IQ is due to shared environment, while genes account for less than 10 percent. In other words, what genes tell you about a child's potential for intelligence is limited to the particular circumstances in which he or she is growing up.

The second problem is even more serious. A developmental outcome that occurs only when a child with a particular set of genes encounters a certain environment is known as a *gene-environment interaction*. Another way to put this idea is that certain genetic characteristics can make a child sensitive to aspects of her experience that wouldn't have any effect on a child of a different genetic background, a theme we will return to later in the book. Such interactions explain the otherwise paradoxical findings that many highly heritable characteristics have increased in the population much faster than biological evolution could explain over the past few decades. Examples range from obesity to intelligence (see chapter 22) to nearsightedness (see *Practical tip: Outdoor play improves vision*, p. 84).

Gene-environment interactions are a problem in this context because researchers assume that the two factors act independently when they calculate those percentages of genetic versus environmental influence that you've seen in the newspaper. But, as we've said, that's rarely the case. Worse still, any interactions that do occur are included in the "genetic" percentage—making the effects of the environment look less important than they really are.

To illustrate these points, let's look at a study of petty criminality in 862 adopted Swedish boys. In this study, either genetics (a criminal parent) or a bad environment (unstable early placement or poor adoptive family) increased the risk of criminality in a child. We would not expect geneticists to identify a "lawbreaking gene," but traits like impulsiveness and aggression are influenced by heredity and can substantially affect a person's odds of breaking the law. Compared to the baseline crime incidence of 2.9 percent for children born to noncriminals and raised in a good environment, the incidence was 12.1 percent for biological children of criminals raised in a good environment and 6.7 percent for biological children of noncriminals raised in a bad environment.

Imagine that genetic and environmental influences were independent of one another. In that case, you could guess the likelihood that a child born to criminal parents *and* raised in a bad environment would commit a crime, simply by adding the two percentages to get 18.8 percent. But the study found something very different. Children with both risk factors, the biological children of criminals raised in a bad environment, had a much higher rate of criminality, 40 percent—more than twice the risk that would have been expected.

From an individual neuron's perspective, it would be hard to distinguish between "genetic" and "environmental" influences.

At the same time, neither factor determined the children's fates. Even under the worst conditions, more than half the children with multiple risk factors turned out to be law-abiding citizens. None of these factors absolutely determines a child's outcome—but they do change the odds.

So the next time you read that intelligence is 60 percent genetic or that researchers have discovered the gene for homosexuality or that children are aggressive only because they've learned the behavior from role models, keep in mind that biology doesn't work that way. Genes and environment are irrevocably entangled throughout your child's life.

PART TWO

GROWING THROUGH A STAGE

Chapter 5

ONCE IN A LIFETIME: SENSITIVE PERIODS

AGES: BIRTH TO FIFTEEN YEARS

Your child's brain is a bit like IKEA furniture: built through a series of steps that normally occur in order. Failing to complete certain steps on schedule—as we certainly have done when assembling a table—can interfere with later steps in the process, usually delaying them but sometimes preventing them from happening at all.

In this chapter, we discuss a special type of development that is central to matching your child's brain to the environment. **Sensitive periods** are times in development when experience has a particularly strong or long-lasting effect on the construction of brain circuitry. Receiving the correct sort of experience during a sensitive period is essential for the maturation of the particular behaviors that rely on that circuitry.

Not all aspects of early development are so demanding. Much of brain maturation occurs without special help. For example, neural circuits in the retina and spinal cord mature according to a set program that is not responsive to experience at all. Other regions—such as the hippocampus and certain parts of the cerebral cortex—are modifiable by experience not during a short period of time but instead throughout life. These brain regions are always able to acquire new information, which helps us continue to adapt to our environment during adulthood.

Sensitive periods for particular functions are special because during these times, the quality of a child's experience can have permanent effects. For example, the brain areas that are specialized for understanding language end up with different connections between neurons depending on whether your baby hears English or Mandarin during the first few years of life (see chapter 6). The brain

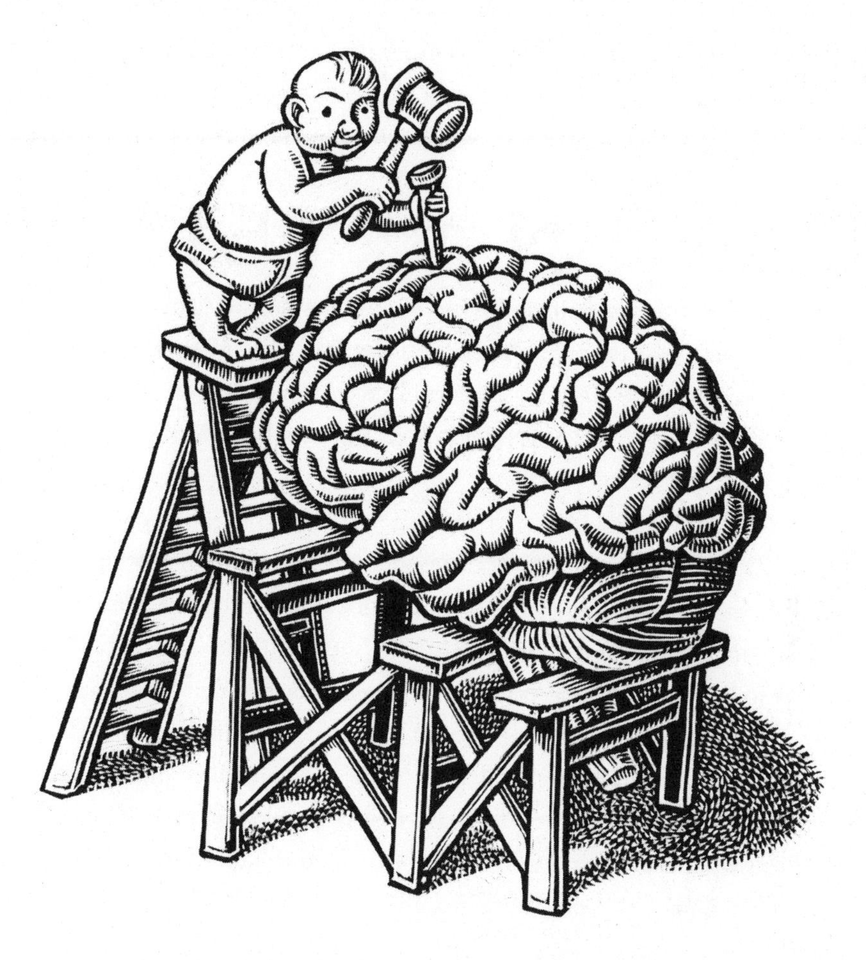

changes that occur in response to this experience make your child an expert at understanding and producing the sounds of his native language.

Later, you may still be able to learn a new language, but you have to work much harder. As an adult, your brain's language areas are no longer under construction and their connections are more difficult to modify. Your sensitive period for language has passed.

Fortunately, as we have said before, experience isn't something that happens passively, even to babies. Your child's brain has definite preferences about what it should learn at various stages of development. The types of experience that can modify a developing neural circuit are determined by predispositions that are built into the brain as a result of our evolutionary history. In short, children actively seek out the experiences they need.

What do we mean by experience? Your child's brain is potentially influenced by any event that can be detected by sensory receptors, transformed into electrical impulses, and transmitted to the brain. (As we will show you in chapters 10–12, all our knowledge of the world comes in the form of these electrical impulses.) Interactions with parents and other caregivers are only part of the rich tapestry of available stimuli. Physical changes in your infant's brain can also occur when she watches the mobile that hangs over her crib, when she sticks her toes in her mouth, and when sirens pass on the street outside. Later, her universe expands to include social interactions with other children, finding her way around the neighborhood, learning to play sports, going to school, and much more. All these experiences leave their traces in her brain, some very long-lasting and others transient.

Because neural circuits mature at different times, there are a variety of sensitive periods in development, each corresponding to a particular brain function. Sensitive periods are most common in infants and toddlers because the brain is undergoing such dramatic growth at this stage, but they can occur at other times as well. Some sensitive periods occur before birth, such as the maturation of the sense of touch based on the baby's experience within his mother's womb (see chapter 11). Many occur soon after birth, as when early interactions with caregivers shape the circuits of the brain that respond to stress (see chapter 26). Other sensitive periods, such as the one for the grammatical aspects of language learning, continue well into childhood and adolescence.

As we described in chapter 2, preprogrammed chemical cues direct axons to their target regions and orchestrate the formation of a large number of synapses. Once those basic elements are in place, experience can influence the further development of the circuit by controlling the activity of those axons and synapses. Synapses that are more effective at activating their target neuron are more likely to be retained and strengthened, through biochemical plasticity pathways in the target cell, while those that are ineffective become weak or disappear. Synaptic activity can also trigger the growth or retraction of axonal or dendritic branches. Cells that fire together, wire together (see chapter 21).

Once these plastic changes have occurred, the brain architecture often becomes more difficult to modify in the future, either because the extra axons and synapses are no longer available or because the biochemical pathways that modify synaptic strength change with age. In this way, the brain uses sensory experience to shape the connections within a neural circuit, pruning away the ones that are unnecessary while maintaining those that are strongest and most active to produce the perceptions and behaviors that are appropriate for that child's individual environment.

Unnecessary synaptic connections are pruned throughout childhood. In the primary visual cortex, the total number of synapses increases rapidly from birth to its peak at eight months old, and then declines slowly to age five, as visual ability is maturing (see chapter 10). The biggest reduction in synapse number in this region happens sometime between ages five and eleven. (We don't know exactly when this change occurs because children ages six to ten have not been studied.) In the frontal cortex, synapse density remains high at least through age seven, falls somewhat by age twelve, and reaches adult levels in the middle teenage years (see chapter 9). It is not clear what happens between ages seven and twelve.

Experience isn't something that happens passively, even to babies. Your child's brain has definite preferences about what it should learn at various stages of development.

Synapse elimination has been studied in much more detail in other primates, and the results are roughly consistent with the sparse human data. In rhesus monkeys, an explosive increase in synapse density in the first few months after birth is followed by an initially gradual and later accelerating decline over the years of childhood. Adult levels of synapse density are reached after puberty. Although the increase is similar across animals, the decline occurs on somewhat different schedules in different individuals, supporting the idea that environmental events influence synapse elimination.

In all areas of the cortex studied in monkeys, synapse development follows a similar time course. It is not clear whether this principle of uniform synapse development also applies to children. Brain scans of developing **gray matter**, where all synapses are found, suggest that frontal regions reach their final volume somewhat later than the visual cortex. However, because of the ages missing from the human synapse counts and variability between individuals, the evidence in support of this position is incomplete. In any event, brain energy measurements

DID YOU KNOW? **BRAIN FOOD**

As kids grow like weeds (and after all, dandelions are weeds), their brains are burning like torches. It's expensive enough to support your mature brain, which uses 17 percent of the body's total energy though it accounts for only 3 percent of the body's weight. But that's nothing compared to the cost of building your child's brain. The brain has nearly reached its full adult size at age seven, but it still contains connections that will be removed later as the child's experiences help sculpt the mature brain. Synapses use most of the brain's energy, so maintaining these extra connections is costly. From ages three to eight, children's brain tissue uses twice as much energy as adult brain tissue. A five-year-old child weighing forty-four pounds (twenty kilograms) requires 860 calories per day, and fully half of that energy goes to the brain.

Researchers examine the brain's energy use with an imaging technique called *positron emission tomography* (a PET scan) that detects radiolabeled glucose, a sugar that is the main fuel for neurons (see figure). (The addition of radioactive atoms allows glucose to be traced through the brain or body.)

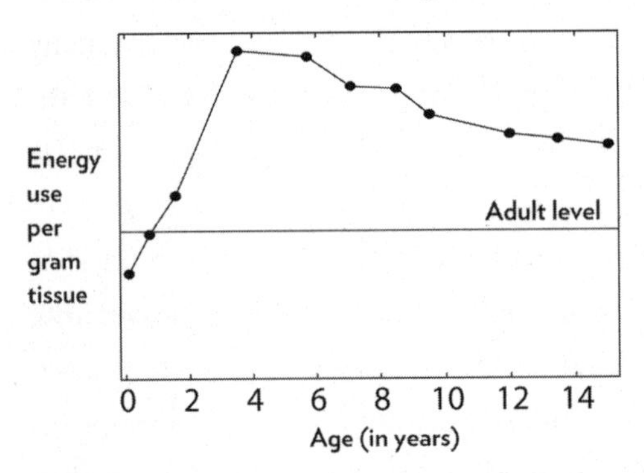

In the first five weeks after birth, energy use is highest in the **somatosensory** and motor cortex, thalamus, brainstem, and cerebellum, the most mature parts of the brain at birth, which are responsible for basic functions like breathing, movement, and the sense of touch. At two or three months, energy use increases in the temporal, parietal, and occipital lobes of the cerebral cortex and the **basal ganglia**, which control vision, spatial reasoning, and action, among other things. From six to twelve months, parts of the frontal cortex increase their energy use, as babies begin to regulate their own behavior. The amount of energy that the brain uses continues to increase until age four and then begins to decline around age nine, reaching adult values sequentially in various areas as they mature, until the pattern becomes fully adultlike in the late teens.

in children suggest that the differences in developmental timing among various cortical areas are relatively minor and that synapse elimination continues throughout childhood (see *Did you know? Brain food*).

To look at the details of how synaptic changes result from experience during a sensitive period, we turn to research in laboratory animals. Barn owls hunt in the dark and must localize sounds accurately to locate their prey. They do this by comparing the difference in the time of sound arrival between the two ears, since a sound coming from the left side will reach the left ear before it reaches the right ear, and vice versa. The more difficult calculation of whether sounds come from above or below is determined from loudness differences created by the shape of the outer ear. An area in the owl's midbrain receives information about discrepancies in timing and loudness and uses it to form a map of where sounds must be coming from. Because the incoming information depends on individual characteristics like head size and ear shape, which change as the animal grows, it can't be specified in advance, so this mapping is learned during development.

The owl's brain learns this map by using visual experience to calibrate the auditory map. To study this process, researchers equip baby owls with prism glasses, which make objects appear to be shifted to one side. At first the animals make a lot of mistakes as they try to move around with the glasses on, but gradually the brain adapts by changing its visual map to reflect the new reality. The auditory space map also shifts in response to prism glasses, even though the auditory information is unchanged.

The shift happens because the neurons that bring in timing and loudness information extend their axon branches to connect with new neurons in a different part of the map. The former connections remain in place, though their synapses are weakened, allowing the owls to return to the old mapping once the prism glasses are removed. This plasticity occurs in a sensitive period, until about seven months of age. In adults, whose sensitive period has ended, it is more difficult to rearrange connections because their axon arbors are limited to a smaller area of the midbrain and thus the wiring is not already in place to carry signals outside the range established in youth.

One of the basic principles of brain development is that the simplest building blocks are finished first. Later, more complex processes build upon earlier ones. For example, the areas of visual cortex that detect edges and shading must become functional before other visual areas can start to interpret these patterns

DID YOU KNOW? THE LIMITS OF BRAIN PLASTICITY

Optimistic popular writers have extolled the wonders of neural plasticity. The idea that experience can produce large changes in the brain is encouraging, as it supports the hope that people can learn and grow throughout life, overcoming obstacles along the way. Stories of untapped potential have a nearly unlimited appeal to the American character. But it's time to step back and take a careful look at the evidence.

Even infants are not blank slates whose brains and behaviors are infinitely modifiable. Before sensory experience can act on a child's brain, the neurons need to be able to talk with each other via synaptic connections. Developmental programs specify particular patterns of connectivity, which are standard for all individuals. Unless there is a genetic error or a developmental accident, the output cells of the eyes will send their axons to the visual areas of the thalamus, which will pass the information along to the primary visual cortex. Axons that carry signals from the touch-sensitive receptors in the fingertips will occupy more space in the somatosensory cortex than axons carrying signals from the less sensitive elbow, and so on.

Under most circumstances, these connection patterns are adaptive, but in unusual cases, this may not be true. In people who cannot see, parts of the visual cortex can be taken over by adjacent regions and used for other functions. Similar types of plasticity allow people to recover from impairments due to strokes by using another part of the brain to compensate for the damaged region. But if the damage is extensive, recovery is likely to be incomplete.

Plasticity outside a sensitive period, if it is possible at all, usually requires more than simple exposure to stimuli. For instance, adults whose vision was impaired by **amblyopia** (also known as lazy eye) can improve their sight after extensive practice on a challenging task, a far cry from the effortless development of the same abilities in normal children. You can change the floor plan of your house after it is complete, but it is much easier to change it during construction.

Retraining the brain in adulthood is possible in some cases, but it is slow and difficult—as it should be. Neural plasticity has costs as well as benefits. Perhaps most important, if routine experience could easily change your brain, you would risk losing hard-won knowledge, abilities, and memories that you acquired earlier in life.

as objects. For this reason, there is not a single sensitive period for vision, but a series of sensitive periods, each requiring experience for the maturation of a different region of the visual brain. If the experience required to complete an early developmental process is not available, the sensitive period is normally extended for a while, resulting in delayed maturation of that brain circuit and all the others that depend on it. Eventually, though, the window of opportunity closes, and any resulting damage may become permanent.

In some cases, higher-level brain areas can compensate for poor development at lower levels, so that adult behavior is relatively unaffected. For example, depth perception can be determined from a variety of visual cues, so people who lack binocular vision due to abnormal visual experience (see chapter 10) often can use other strategies to determine depth accurately.

As we have already said, learning language requires experience during a sensitive period. In extreme cases, children who grow up in a poor-quality language environment can fall progressively further behind as development continues. But in normal circumstances, babies are sponges for language. You don't need to train your baby to imitate your voice instead of the sounds of the family car because her brain areas for language are best activated by speech sounds and because language acquisition, like so many other types of learning, is most effectively driven by social interactions. In the next chapter, we will consider language further as a well-studied example of a sensitive period.

Chapter 6

BORN LINGUISTS

AGES: BIRTH TO EIGHT YEARS

Complex skills require deep foundations. Babies start to learn language a long time before they are able to speak, preferentially focusing their attention on speech from birth—or even earlier, as hearing becomes functional during the third trimester of pregnancy (see chapter 11). Because babies do not have the motor abilities to express all the knowledge that they have obtained, though, you may not realize how much language they understand at a given age.

Newborn babies already prefer their mother's voice over other female voices, their native language over other languages, and speech over other sounds that have the same acoustic properties, including speech played backward. They can also detect a variety of vocal cues, including acoustic characteristics, stress patterns, and the rhythms of different languages. From early in life, your infant absorbs the huge amounts of information that will make him an expert in his native language, learning about its cadences, its sounds, the structures of its words, and the grammar of its sentences. As we discussed in chapter 3, most adults instinctively speak to infants in motherese, which is slower than normal language and contains exaggerated versions of consonant and vowel sounds.

Young infants can distinguish and categorize the sounds of all languages of the world, though adults often confuse the sounds of a foreign language. For example, the *r* and *l* of English sound the same to Japanese adults, but different to Japanese infants. As they acquire experience with speech, babies begin to specialize in the sounds (called *phonemes*) of their own language (or languages). By six months of age (for vowels) or ten months (for consonants), babies become better at identifying the phonemes of their native language and worse at identifying the phonemes of other languages. In other words, experience with language shapes the categories into which babies place sounds, determining which variations in sound characteristics are meaningful (reflecting different phonemes) and which should be ignored (reflecting different speakers or other unimportant variations).

As we would expect, their neural activity reflects this phoneme learning. In older infants, the patterns of electrical signals in the brain recorded from electrodes on the scalp, termed *event-related potentials*, show that babies distinguish between a pair of sounds from the native language, while failing to distinguish two confusable foreign sounds. In younger infants, event-related potential patterns distinguish both foreign- and native-language sound pairs. This brain specialization is important for future language learning. Babies whose brains discriminate native sounds well (and foreign sounds poorly) at seven and a half months go on to learn language earlier than babies who show the less mature pattern of distinguishing all sounds equally well. The more discriminating babies learn words more quickly, produce more words and more complex sentences at twenty-four months, and produce longer phrases at thirty months than the less discriminating babies. So even though your baby isn't talking back, he is absorbing the patterns of your talk.

Social interaction is one cue that babies use to determine which sounds they should be learning. Nine-month-old infants who hear a brief tape recording or video of someone speaking a new language do not learn its sounds, but the same

amount of speech from a live person is sufficient to allow the babies to discriminate phonemes in the new language. (Under some circumstances, babies can learn from tape or video, but it takes longer than learning from a live person.) Indeed, certain measures of social interaction with a language teacher (including a parent) predict how well individual infants will remember the sounds of the new language. The preference for social interaction may be part of the reason that autistic children (see chapter 27), who do not interact well with other people (and do not prefer the sounds of motherese), have difficulty learning language.

The timing of speech production is determined by maturation of the brain regions that control movement. Forming understandable sounds requires considerable fine motor control and apparently a lot of practice. Babies first attempt to talk at around two months, when they begin cooing vowels, the least complicated speech sounds to produce. Some consonant sounds follow around five months, when babbling begins. Early babbling sounds the same in all babies, regardless of their native language. Around the end of the first year, babbling starts to include language-specific phonemes.

Responding with a comment or a touch to your baby's best attempts to communicate seems to encourage continued efforts to improve these skills.

Word learning also starts long before babies can produce words of their own. Six-month-old infants know their own names and will look at a picture of their *mommy* or *daddy* when they hear the word. As we discussed in chapter 1, infants can listen to a string of nonsense syllables and determine which of them are most commonly heard together as "words." They apply this talent to identifying words in normal speech, where words tend to run together without pauses. (To understand this phenomenon, think of the way a foreign language sounds; you can't guess where one word ends and the next begins.) Later, their brains learn about the regularities of sentence structure that constitute the rules of grammar in their native language. By nine months, familiar and unfamiliar words trigger noticeably different event-related potentials. By the first half of baby's second year, these potentials are different for words whose meaning the child does or doesn't understand. Babies' brains also respond differently to made-up words depending on whether or not they obey the rules for which syllable should be stressed in the baby's native language. Stress patterns appear to be another tool that babies use to determine which groups of sounds are words.

In the second year, as children learn more words and become able to say many of them, they become better at distinguishing similar words, like *bear* and *pear*. Babies at fourteen months will direct their gaze toward an object even when its name is mispronounced, suggesting that their brain does not yet represent the sounds in known words with complete accuracy. Similarly, at this age, brain activity does not distinguish between familiar words and similar-sounding nonsense words. This changes at around twenty months. The relationship between learning words and learning sounds seems to be bidirectional, so that learning sounds makes it easier to learn words, but learning more words also helps babies improve their ability to distinguish sounds.

Sentences add new layers of complexity to language learning. Again, children can comprehend sentences and grammatical connecting words before they're able to use them in speech. To understand a sentence, your child must know not only the meanings of the individual words (called *semantic information*) but also how they relate to each other within the sentence (*syntactic information*). The brain represents these two types of information separately.

For almost everyone (excepting some left-handers), the left hemisphere is dominant for language production. Similar regions in the right hemisphere are responsible for *prosody*, the tone and rhythm of speech that conveys much of its emotional content. (For example, prosody tells you when someone is being sarcastic or making a joke.) Laterality of language representations seems to be part of the basic pattern of brain connections laid down by genes before sensory experience becomes effective (see chapter 2) because it is apparent by two or three months of age and even occurs in deaf infants. If the dominant speech regions are damaged in childhood, though, especially before the age of five, the other side of the brain can take over their function, leaving language skills relatively normal. If the same damage occurs after puberty, it severely impairs communication abilities.

When we hear something that sounds "wrong," event-potentials in our brains reveal whether we're reacting to syntactic or semantic violations. "The boy walked down the flower" is an example of a semantic violation, while "The boy walk

PRACTICAL TIP: TEACH FOREIGN LANGUAGES EARLY IN LIFE

From the perspective of neuroscience, it's absurd to wait until high school to begin studying a foreign language. By adolescence, students must work much harder to learn a new language, and most of them will never master it completely. If you want your child to speak another language fluently, by far the best approach is to start early in life.

In one study, researchers tested the English grammar proficiency of Chinese or Korean immigrants who had arrived in the U.S. at various ages and stayed at least five years. The test required participants to identify whether there were grammatical errors in sentences like "Tom is reading book in the bathtub" or "The man climbed the ladder up carefully." The test was simple enough that native English speakers could ace it by the age of six, but the immigrants who began learning English after age seventeen missed many of these simple questions. Only people who came to the U.S. before age seven performed at the level of native speakers. Everyone in the group who arrived at eight to ten years of age did a bit worse, and those who arrived at eleven to fifteen were still less proficient.

Between ages eight and fifteen, researchers found a strong relationship between age of exposure and performance on the test. But in adulthood, individual variability in performance was not connected to age. No matter whether they'd started learning English at eighteen or forty, few adults learned perfectly. (Some later researchers found that language learning in adulthood also declines with age—that is, young adults learn better than older adults—but everyone agrees that young children learn better than older people.)

The take-home message for parents and schools is clear: take advantage of young children's superior language learning abilities by beginning instruction in elementary school or earlier. When it comes to language, there's no substitute for an early start.

down the road" is syntactic. In small children, these mistake-detection responses develop slowly, starting as children transition from two-word phrases to their first full sentences, around thirty months of age. Brain responses gradually become faster and more precisely localized through childhood and into the early teens.

There seem to be at least two sensitive periods for language learning. We already discussed the sensitive period for phonemes, in the first year or two of life, when babies' brains become specialized for representing the sounds of their native language(s). There is also a sensitive period for learning about grammar. Children's ability to acquire syntax rules declines gradually after age eight, and adults are worse than children at learning languages (see *Practical tip: Teach foreign languages early in life*).

Some adults manage to learn a second language to a high level of proficiency. Most of us, though, no matter how hard we study in adulthood, will always have a foreign accent and make minor grammatical errors. In contrast, there does not appear to be a sensitive period for semantic learning, as new vocabulary words can be acquired equally well at any age. The event-related potential signal for semantic violations looks the same for both native and second languages, even in people who learned their second language late in life.

Children can learn more than one native language if they are exposed to both languages early enough, but their brains appear to represent the languages at least somewhat separately. Bilingual children reach language milestones at the same age and have the same risk of language impairment as monolingual children, though the details of their language development are somewhat different. So if your household is bilingual, the research indicates that this is not a disadvantage for your child's language learning. (Indeed, it may be an advantage for cognitive development; see *Practical tip: Learning two languages improves cognitive control*, p. 118.) Learning a second language also changes the brain. A region in the left inferior parietal cortex is larger in people who speak more than one language, and it is largest in those who learned the second language when they were young or speak it fluently.

Infants quickly learn to identify different languages by their rhythms, their characteristic phonemes, and other cues. Bilingual children do sometimes mix languages in their speech, but they seem to do so for the same reasons and in the same situations as adult bilinguals, for instance, substituting a word from one language when they don't know the word for that concept in the other one. Though bilingual children have a smaller vocabulary in a particular language than monolingual children of the same age, bilingual children know more words in total if you count both languages.

Children who hear more words while interacting with their parents in the first two years of life learn language faster than children who hear fewer words. These differences in home environments tend to fall along socioeconomic class lines. In one study, the poorest children heard 600 words an hour, working-class children heard 1,200 words, and children of professionals heard 2,100 words. These major differences in children's language environment correlate with their later language development and IQ scores—though the finding that highly verbal parents raise highly verbal children may be partly due to genetic factors or the many other advantages of growing up in a professional household (see chapter 30).

Later research has shown that you can improve your children's language skills by responding rapidly to their vocalizations, mimicking the turn-taking of conversation even before your baby is capable of forming words. Responding with a comment or a touch to your baby's best attempts to communicate seems to encourage continued efforts to improve these skills. So talk to your baby and put up a good show of understanding what she's saying. It's fun for both of you, and it will help her language skills to develop more quickly.

Chapter 7

BEAUTIFUL DREAMER

AGES: BIRTH TO NINE YEARS

Sleep appears to be simple, but it is composed of many brain mechanisms working together—mostly seamlessly. In babies and young children, these brain functions mature at different times. The complex abilities involved in sleep become apparent in stages as your child grows. The intense need for sleep early in life may be related to its importance in facilitating learning.

The first function to appear, well before birth, is the internal **circadian** (Latin *circa dies*, meaning "approximately a day") **rhythm**. This clock can run for many days without external instruction, providing our brains and bodies with cues about our daily activities even if we can't see the sun. The brain can generate an approximately twenty-four-hour rhythm without light, using a complex signaling clockwork made of genes and proteins. This clockwork's output is used by other brain regions and organs to set their own daily rhythms, for hunger, bowel movements, body temperature, liver activity, and stress hormone secretion.

This daily rhythm of our brains and bodies is driven by the **suprachiasmatic nucleus**, a dab of tissue containing fewer than ten thousand neurons that sits over the **optic chiasm**, where the optic nerves meet and cross on their way toward the brain. The suprachiasmatic nucleus gets its signals from ganglion cells in the retina that are dedicated to transmitting information about light levels in the world. These ganglion cells, which make a pigment protein called **melanopsin**, convert light to impulses that travel along the optic nerve to the suprachiasmatic nucleus. In this way, the brain knows when it is day and when it is night.

The fetus has a suprachiasmatic nucleus at eighteen weeks of gestation. Several

weeks later, circadian cycles are found in the fetus's heartbeat and breathing. This rhythm is probably driven by day-night signals from the mother, such as the rhythmic release of three hormones: **corticotropin-releasing hormone (CRH)**, **cortisol**, and the sleep-inducing signal *melatonin*.

Once your baby is born, that rhythm is suddenly lost. As any new parent can wearily tell you, newborns have highly irregular sleep patterns, though it is possible to drive the rhythm a bit through feeding times. Starting around three months of age, your infant's sleep-wake patterns start to be influenced by cues such as the timing of feedings and nighttime routine. So you can make the baby's pattern regular by providing a set daily routine. Even so, for the first few weeks after birth, there is almost no day-to-day pattern. The sleep-wake cycle in infants typically lasts about fifty to sixty minutes, with no relationship to time of day. Later, this cycle of sleep fades into an *ultradian* (meaning "less than a day long") cycle of greater and lesser alertness, one that has a rhythm of about an hour through age three and lengthens by age five to ninety minutes, the cycle that continues for the remainder of life.

Infants spend about two thirds of their time sleeping, about half in active sleep that is similar to adult rapid-eye-movement (REM) sleep. In adult REM sleep, muscles under voluntary control shut down except for the eyes, a phenomenon that prevents us from acting out our dreams. In contrast, infants are hardly mobile and not in a position to act out much of anything. In the face of such relative safety, your infant can move a lot during active sleep: he can make sucking motions, twitch, smile, frown, and even move his limbs. But it is unlikely that he is dreaming about movement or action, if indeed he dreams at all (see *Did you know? What children dream about*, p. 60).

With or without dreaming, babies' active sleep may serve essential functions. As we saw in chapter 5, early development is a period of massive growth and pruning back of neuronal connections. Neuroscientists can observe these changes most clearly in animals. In cats, kittenhood is a time when dendrites in the neocortex are remodeled in response to changes in visual experience—and conversely, to visual deprivation. Sleep enhances this remodeling process, a necessary part of development. Conversely, sleep loss reduces the changeability of dendrite structure, a major component of neural plasticity. Research suggests that this is true in both adult and juvenile mammals. One possible consequence of all this remodeling is that sleep may help to consolidate some kinds of learning, including the transfer of information from short-term to long-term memory (see chapter 21).

Over the next six months, as the baby encounters daily light and dark cycles for the first time, the circadian rhythm is gradually regained. The first sign of progress is a slight drop in core body temperature in the early morning every day. By three months, about two thirds of children sleep at least five hours at night.

The conscious, awake state depends on columns of neurons deep in the brainstem called the *reticular formation*. The reticular formation is roused to activity when we are awake by mechanisms that are not fully understood, but involve neurotransmitters that are secreted in the brain's core, such as **acetylcholine** and **norepinephrine** (also called *noradrenaline*).

REM sleep is controlled by neurons in the **pons**. These neurons send axons—and commands—both forward and downward. Descending connections prevent motor neurons from firing and therefore from causing muscle contraction, using several as-yet unidentified neurotransmitter pathways. The function of the forward-projecting connections is unknown, but they may drive plasticity—and perhaps dream activity. During non-REM sleep, movement is still possible, but sensory input does not get through very well, especially during deep non-REM sleep.

As babies grow, sleep changes. The amount a baby sleeps declines gradually, reaching twelve hours per day by age two. At the same time, REM sleep diminishes dramatically, as do nighttime melatonin levels. By age three, children spend just one fifth of sleep time in REM sleep, the same proportion as teens and adults. By age five, sleepers alternate between non-REM and REM sleep over a ninety-minute period, the same duration as the adult sleep cycle.

The development of sleep does not always go smoothly. During sleep, many events need to be suppressed to keep the sleeper safe, whether animal or human. The sleeper doesn't urinate. The sleeper doesn't act out dreams by walking around or making noise, thereby attracting predators. In children, these safety mechanisms are still under construction, as you may have noticed if your child has not yet acquired them. Before the age of six, one child in three experiences interruptions called *parasomnias*, which include a suite of problems such as sleepwalking, bedwetting, and night terrors. If parasomnias do occur, they begin between three and six and are largely resolved by the onset of puberty. Parasomnias usually happen during the first few hours of the night, during the deepest sleep just before the evening's first bout of REM sleep.

Watch for drowsiness and act quickly: put the baby down to sleep right away. Babies cycle in alertness, just as adults do, but more quickly.

One parasomnia can be particularly upsetting to a parent: night terrors, which occur in 1 to 6 percent of children age three or older. In younger children, they can occur at least once a week. Night terrors consist of waking, typically from deep non-REM sleep, with expressions of fear, often including screaming.

PRACTICAL TIP: HOW TO GET YOUR BABY TO SLEEP

The amount and type of sleep that a child needs is programmed over the course of normal development, without much input from the environment. Triggering sleep is a somewhat different story, one that depends on learning mechanisms.

Babies' sleep can be made more regular by establishing a regular feeding schedule (see main text). We should warn you that one old-fashioned trick for inducing drowsiness has been shown to backfire. The trick is to add a small amount of alcohol to mother's milk, for instance, as would occur if Mom has a glass of wine. It is true that this can slightly shorten the time until sleep begins, by about fifteen minutes. The total amount of sleep over the next three and half hours is reduced by an even greater amount, though, suggesting that this is a bad strategy. A better strategy is to watch for drowsiness and act quickly: put the baby down to sleep right away. Babies cycle in alertness, just as adults do, but more quickly.

For bedtime, small children learn quickly to associate particular cues with sleep. Like other associations, children can learn from a single example—or a single exception. Once formed, preferred bedtime habits are hard to break. For example, if children become used to having a parent present at the onset of sleep, that becomes a requirement. It's better to leave the room so the child associates falling asleep with having the parent absent—which will turn out to be a boon later. In general, establishing a bedtime routine, including toothbrushing, stories, and winding down of attention paid to the child, provides a familiar landing procedure. Consistency is essential. For more information on this subject, an excellent resource is *Healthy Sleep Habits, Happy Child* by Marc Weissbluth.

The child is inconsolable and takes from five minutes to half an hour to settle down. In the morning, he does not remember anything.

Night terrors are not simply nightmares. They occur at an age when children's dreams do not include fear, or for that matter any other emotion (see *Did you know? What children dream about*, p. 60). In night terrors, the child is fearful but cannot report any specific events. One component could be lack of regulation

DID YOU KNOW? WHAT CHILDREN DREAM ABOUT

Most of us are familiar with the claim, popularized by Sigmund Freud and Carl Jung, that dreams symbolize hidden desires and fears. Perhaps in part because of this culturally driven expectation, we often assign meaning to our dreams after we wake up, making our reports unreliable.

Your child may also experience a version of this coaching if you ever ask what he dreamed about or wish him good dreams. Your wish is harmless, but it also means that you're inadvertently encouraging storytelling that might not match what was actually dreamed. Unless adults practice recalling their dreams, most of a night's dreams are forgotten by morning. The same is likely to be true for children.

If you ask adults, they report having a dream 60–90 percent of the time after being woken up from REM sleep, and 25–50 percent of the time from non-REM sleep. They can also report what they were dreaming about. You can take this approach with children too: wake them up and ask what they were dreaming about (or what was happening, if they are very young and do not know what a dream is). If you are unwilling to wake your sleeping child in the name of science, you don't have to. The sleep researcher David Foulkes and his collaborators stayed up with children between the ages of three and twelve, either in a laboratory or at the children's homes. They woke the children in the night and then asked what was happening just before they woke up.

At early ages, children's dreaming was simple and rare. Only 15 percent of three- and four-year-olds reported any dream events after being awoken from REM sleep, and there were no dreams at all during non-REM sleep. The dreams that preschoolers did report were static and often involved animals: "a chicken eating corn" or "a dog standing." The children themselves typically appeared as passive agents: "I was sleeping in a bathtub." "I was thinking about eating." At this age, dreams also lack social interactions or feeling. Preschoolers do not report fear in dreams, nor do they report aggression or misfortunes. This contrasts with their waking lives, in which they can describe people, animals, objects, and events around them.

At later ages, dreams take on more complex qualities. Around age six,

dreams become more frequent and acquire active qualities and continuity of events. By eight or nine, dreams are reported as frequently as they are in adults, have complex narratives, and feature the dreamer as an active participant. Dreams also start to include thoughts and feelings.

The static dreams of preschoolers occur when their visual-spatial skills are not fully developed. For instance, children who report more dreams are also better at re-creating pictures of red and white patterns using blocks. At this age, children are also less able to imagine what an object looks like when viewed from a different angle. Such skills depend on the parietal lobe, which sits between neocortical regions that represent vision and space. The parietal lobe does not become fully myelinated until age seven, suggesting that at earlier ages, children may dream in static images because their brains are not fully competent to process movement.

What does this process tell us about child development? One potential answer is that dreams reveal the patterns of brain activity that are possible in the absence of external stimuli. In this respect, they may give you a window into the developing conscious mind of your child.

---

of brain structures that handle strong negative emotions, such as the amygdala. The amygdala regulates the sympathetic nervous system to mobilize the brain and body to fight or run away. In adults, this system can be suppressed by other regions of the brain such as the hippocampus and neocortex. In children, who have less developed control over their emotional responses (see chapter 18), such suppression may not be mature enough to be effective, especially during deep sleep.

Night terrors may be triggered by sudden waking. In one study, eighty-four preschoolers who sleepwalked or suffered night terrors were observed while they slept. More than half were found to have disordered breathing, such as obstructive sleep apnea. In this disorder, when respiratory centers in the brainstem receive a signal that breathing has stopped, the sufferer wakes up suddenly, gasping. In children, sleep apnea has several causes, including being overweight and having enlarged tonsils. Remarkably, all the children with sleep apnea who underwent tonsillectomies in that study were cured of their night terrors.

In night terrors and sleepwalking, sleep mechanisms do not fully suppress behaviors that usually occur in waking life. In the other direction, sleep phenomena can also appear unbidden in the daytime. Examples include narcolepsy and cataplexy, in which an awake person suddenly falls asleep or loses conscious motor control. An upsetting example is sleep paralysis, in which a person wakes up but cannot move. In this case, touching the person is usually enough to end the paralysis. Don't be reluctant to do this—you will be rescuing your child from anxiety.

Another characteristic of children's sleep is the need for naps, which is driven by a slow cycle that spans the entire day. In both adults and children, alertness and sleepiness are cyclical. Part of the cycle includes an afternoon lull, followed by a second wind. Until age five or six, the lull is low enough to require an afternoon nap. After that age, children are able to stay up all day, as adults can.

This is not to say that staying awake all day is the best strategy for grown-ups. Low afternoon alertness in adults may be a remnant of our need for a nap. In one experiment, college students were required to focus on the center of a screen when the letter *T* or *L* flashed, followed after a brief pause by some diagonal bars shown elsewhere on the screen. Students tested early in the day identified both the letter and the orientation of the bars even if they were flashed in quick succession. Later in the day, they needed the letters to be spaced at longer intervals for successful identification. This slowing of perceptual capacities was prevented if the students were allowed to take a brief nap. Naps not only keep toddlers from getting cranky but may also help adults' mental performance.

So when you watch your baby or child sleeping, be aware that her brain is far from idle. Her brain is fulfilling a well-choreographed process that is coming along nicely without any special effort by her or you. While the sugarplums in her head might not be dancing, bigger things are changing—including events that may restore and rewire the developing brain.

Chapter 8

IT'S A GIRL! GENDER DIFFERENCES

AGES: BIRTH TO EIGHTEEN YEARS

Three-year-olds take gender roles as seriously as drag queens do. One of our colleagues, who was dedicated to freeing her kids from traditional gender expectations, bought a doll for her son and trucks for her daughter. She gave up her quest after she found the boy using the doll to pound in a nail and the girl pretending that the trucks were talking to each other.

Many puzzled parents have wondered where this highly stereotyped behavior comes from, especially in households where Mom wouldn't be caught dead in a pink frilly dress and Dad would rather cook dinner than watch sports. All over the world, a phase of intense adherence to a sex role seems to be important for the development of a solid gender identity. This stubbornness reminds us of the early stage of grammatical learning, another area where young kids apply newly learned rules more broadly than necessary ("That hurted my foots" instead of "That hurt my feet").

In light of their behavioral differences, you'd probably imagine that the brains of little boys and girls are distinct in many important ways. Because of our society's intense interest in sex differences, researchers have done many thousands of studies of this topic, and journalists have been eager to publicize them. This literature is vast and variable, so as we evaluated it, where possible we relied on meta-analysis (see glossary) to evaluate the findings. From careful review of such papers, a few important patterns emerge.

When we evaluate reports of sex differences, it's important to pay attention to the size of the effects. Most gender differences are too small to matter in any practical way, and a minority of differences are important when comparing groups. But only a few tell us anything significant about individuals. For instance, girls—on average—are more likely to hear a relatively quiet sound. But it would be impossible to guess whether a particular child is male or female by knowing that child's hearing ability, because all possible scores are found in both boys and girls. And what's more, for nearly all sex differences, the differences among individual girls or among individual boys are much larger than the average differences between the sexes, with a few important exceptions.

What do we mean when we say that a gender difference is small or large? Let us be technical for a moment. Scientists often measure the size of a difference between two groups by calculating a statistic called *d-prime (d')* or **effect size**, defined as the difference between groups divided by the standard deviation, a measure of variability, of one or both groups. If there is no difference, the d' is zero. The d' gets bigger as the size of the difference in average scores between the groups increases, relative to the range of scores within each group.

This idea is easier to explain in pictures than in words. The figure below shows the differences between groups that correspond to typical d' values. The

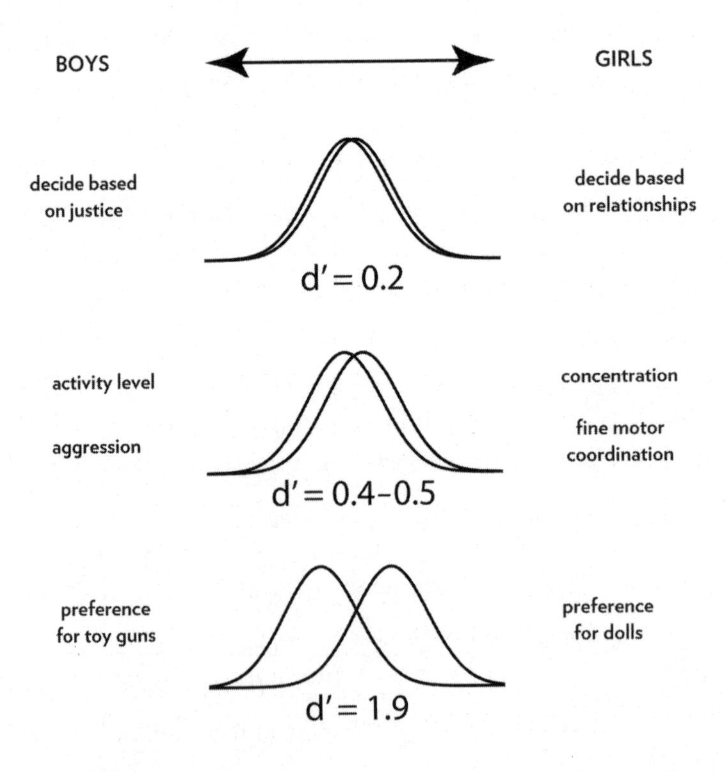

horizontal axis represents the possible scores, while the height of the curve represents the number of people in the population who get a particular score. From top to bottom, these differences would be considered small, medium, and very large.

Let's consider some specific examples. For gender differences in adult height in the U.S., the left curve in the bottom panel (d' = 1.9) would represent women, and the right curve would represent men. The horizontal axis would show heights from short (left) to tall (right), with the peak of the female curve at 5 feet 3.8 inches, the average height for women, and the peak of the male curve at 5 feet 9.4 inches, the average height for men. A man of average height is taller than 92 percent of women. In the research literature, a value of d' that is at least 0.8 is considered large, so this would be a very large difference.

At the other extreme, let's take as our example a small difference that we've already mentioned: hearing. Several authors have recently argued in favor of single-sex education based in part on the idea that girls have more sensitive hearing than boys and therefore respond best to teachers speaking quietly. For hearing sensitivity, the left curve in the top panel (d' = 0.2) would represent boys, and the right curve would represent girls. Because the difference between the two groups is small, as you can see, the two curves overlap substantially.

The individual differences in hearing within each sex are much larger than the differences between boys and girls. And given that many boys have sharper hearing than many girls, it doesn't make sense to argue for sex segregation on these grounds. If you think sensitive hearing affects the way people learn, you should separate them based on their hearing, not their gender—the two are not the same.

Only a few gender differences are big enough to predict individual behavior. The largest known behavioral difference at any age is toy preferences in three-year-olds. Parents who try to keep their sons from playing with toy guns often discover that any stick—or, in a pinch, even a doll—can be converted to a weapon in a boy's imagination. Given the choice between a boy-typical toy like a car and a girl-typical toy like a tea set, at age three children differ in their choice of boyish toys with a d' of 1.9, a difference corresponding to the bottom panel of the figure. This means that you can do quite well at guessing the sex of a young child based on his or her choice of toys, as 97 percent of boys are more likely to play with male-typical toys than an average girl. Because play helps children learn and practice a variety of skills, sex differences in how children spend their time can influence which abilities they carry through life (see *Practical tip: Broadening your child's abilities*, p. 68).

The emergence of toy preferences is an early stage in the development of *gender identity*, defined as your child's self-identification as male or female. Gender-influenced toy preferences are seen across cultures, beginning around one year of age. Even babies have some understanding of gender (see chapter 1), but only a few two-year-old children can accurately state whether they are boys or girls or reliably distinguish men from women in pictures. Most children—again across cultures—reach this milestone by two and a half, and almost all children get there by age three. Children who have reached this milestone are less likely to choose the "wrong" toy than children who have not.

Toy preferences almost certainly have an innate basis (though they are also influenced by culture). One clue is that male monkeys prefer to play with trucks, while female monkeys prefer dolls. Another clue is that boy-typical toy preferences are more common in girls with a syndrome called **congenital** *adrenal hyperplasia* or *CAH*. Due to a genetic defect in adrenal hormone synthesis, CAH girls are exposed to an excess of **testosterone** and other androgens, masculinizing their brains and to some extent their bodies in utero. Because this hormonal malfunction can be treated starting at birth, CAH girls offer an opportunity to look at the effects of prenatal exposure to male hormones on later behavior.

As they get older, girls tend to become more flexible in their toy preferences. By age five, nearly half will pick a boy-typical toy when offered a choice. Boys, on the other hand, continue to refuse girl-typical toys, most likely because the social penalty for acting like a girl is very steep. Both peers and parents—especially fathers—actively discourage boys from playing with girl toys.

Some parents are concerned that allowing their son to play with girl toys will lead him to be gay in adulthood, but this worry confuses correlation with causation. Whether parents encourage or approve of their son's habits is irrelevant to his sexual orientation later in life. It is true that about half of the boys who prefer girl toys do grow up to be gay—and also true that many of them do not. (Tomboy girls, on the other hand, rarely turn out to be lesbians.) Playing with dolls doesn't cause boys to become gay, though. The most likely explanation is that playing with girl toys and adult homosexuality both result from earlier influences on some boys' brains, perhaps due to prenatal experiences or genetics. Psychiatric treatment aimed at encouraging boyish behavior has no effect on

adult sexual orientation, but the father who tries to discourage girlish behavior might be opening a rift with his son. By the time you can observe the behavior, the outcome is out of your control, so you might as well get comfortable with it.

You've probably noticed two other sex differences in young children's behavior. Boys are significantly more active and more physically aggressive than girls. These differences are medium sized, with a d' of 0.5, corresponding to the second panel of the figure on p. 64. That means an average boy is more active and more physically aggressive than 69 percent of girls, which does not predict individual behavior very well, but does make groups of boys obviously different from groups of girls. These differences are also probably influenced by the action of hormones on the brain, not just by culture. Juvenile male monkeys show more rough-and-tumble play than female monkeys, and this behavior can be modified by hormone treatment. Similarly, CAH girls are more aggressive and more active than other girls, again suggesting an early hormonal influence on this behavior.

PRACTICAL TIP: **BROADENING YOUR CHILD'S ABILITIES**

Children's play may affect their later behavior and interests. You can't force boys to behave like girls or vice versa, but by taking your children's natural inclinations into account, you can help them to practice skills that they might not find on their own. You don't know what the future holds, and we figure that you can't go wrong by increasing the number of options available to them in adulthood.

One of the largest adult sex differences is that males are better at mentally rotating objects through space. (This ability affects the way we think about directions, as well as some practical skills like moving a couch through a doorway.) This pattern emerges early in life and is then modified by later experience. Many male infants at three to five months can recognize rotated objects, while few female infants of the same age can do so. Otherwise infants show no sex differences in their understanding of the behavior of objects (see chapter 1). In elementary school, the gap in mental rotation ability is small, but it continues to widen as children mature, reaching a d′ of 0.66 to 0.94 (depending on how the test is scored) for adults, meaning that the average man performs better than 75 to 83 percent of women. Performance on mental rotation tests predicts performance on the math part of the SAT (originally Scholastic Aptitude Test, later renamed Scholastic Assessment Test I) in both male and female high school students and likely contributes to sex differences in map reading and navigation ability.

It makes sense that different styles of play might improve different skills. Exploring physical objects and their interactions is an important component of boys' play. As they build towers of blocks and knock them down, wrestle, play catch, or ride bikes around the neighborhood, boys are learning about the rules of the physical world. As girls play with dolls and dollhouses, they are practicing nurturing and fine motor control skills. Girls also talk with each other during play more than boys do, which may help girls to become more fluent and have larger expressive vocabularies by the time they start school.

Boy-style play develops spatial skills in all brains. Boys raised in deprived conditions don't show an advantage over girls in their spatial abilities. In one study, boys from families with low socioeconomic status (SES; see chapter 30) scored lower on a mental rotation test than boys from medium- or high-

SES families and performed no better than girls of any SES. Boys from such families may not get the play experiences required to develop their object manipulation skills. Video games involving navigation or other spatial tasks help boys and girls learn to visualize and rotate objects. Some studies find that these training effects are especially large in girls. Playing sports may also be helpful. College athletes of both sexes have an advantage over non-athletes in mental rotation tasks and other spatial skills, though this may be because people with good spatial abilities are more likely to play sports. Researchers have not yet demonstrated that these experiences lead to real-world improvements in spatial skills, but that will be the next step.

How can parents help all their children develop a broad range of abilities? Encouraging girls to play video games could improve their spatial reasoning (as well as their comfort with computers). We also suggest getting girls involved in sports (see chapter 15) when they're young, as self-consciousness may inhibit teenage girls from wanting to learn new physical skills.

Parents can help boys to develop better language skills by talking and reading to their sons, starting in infancy (see chapter 6). Boys may also benefit from extra help with phonological awareness in the preschool years, which parents can provide by discussing which letters make which sounds as they read. Similarly, you can take advantage of young boys' attraction to the computer to encourage them to write stories onscreen or choose books about fighter pilots or dinosaurs to engage their interest in reading. You can find many other suggestions for helping girls and boys grow into well-rounded adults in the neuroscientist Lise Eliot's book *Pink Brain, Blue Brain*.

Perhaps as a consequence of these behavioral differences, starting as early as two or three years of age, children prefer to play with other children of the same sex. These segregated play groups persist throughout elementary school. This pattern is seen across many societies, from agricultural villages to big cities. It does not seem to depend on whether or not the adults in the society have strongly segregated sex roles, though cultural factors can modify the pattern's expression somewhat. It even occurs in monkeys and apes. If there are only a few children available, boys and girls will play together, but when possible, they usually split the group by gender.

This behavior reinforces gender norms as children are learning about their gender identities. Social pressure from other children to conform to gender norms is particularly strong from age four to eight, perhaps because children's early concepts of sex roles (and many other rules of society) tend to be written in black and white, with a more flexible understanding emerging later in development. We know a neuroscientist couple whose son was best friends with a girl throughout the preschool years. When he was six, his male friends at summer camp let him know that playing with girls was unacceptable. Now their son will only see his female friend inside the house, with everyone involved sworn to secrecy so the other boys won't find out. Single-sex groups often act in such a way to communicate and enforce gender-specific behaviors.

Parents who try to keep their sons from playing with toy guns often discover that any stick—or, in a pinch, even a doll—can be converted to a weapon in a boy's imagination.

Sex differences in behavior give girls a medium-sized advantage over boys in the classroom, where girls get better grades in high school and college. Girls' brains mature earlier than boys' brains, with the peak volumes of most brain structures occurring one to three years sooner in girls. Girls are moderately better at inhibitory control (d' of 0.4)—that is, sitting still and concentrating on their task—so the classroom culture is more friendly to girls. On average, girls are a bit more advanced in some areas of verbal development when starting school. Boys lag at fine motor coordination (d' of 0.6), giving them a moderate disadvantage in the ability to write letters, the largest sex difference in academic performance as school begins. These gaps persist through high school, with boys continuing to score lower on tests of both reading and writing.

But let's put these gender comparisons in perspective. All these differences have smaller effects than the difference between living in a middle-class neighborhood with good schools (judged by their average test scores) and living in a low-income neighborhood with poor schools. First graders from poor areas score lower than their middle-class counterparts (d' greater than 1.1) in both reading and mathematics performance, and those gaps too typically widen with age (see chapter 30).

Like other gender differences, gender-related differences in education may be modified by experience. Girls have recently caught up with boys in academic areas where they were lagging just a decade or two ago. In the U.S., there are no remaining gender differences in average performance on mathematical achievement tests through high school. In addition, women are now more likely than men to attend and complete college. In the U.S., there are 185 women for every 100 men with college degrees at age twenty-two. Because men take longer than women to graduate, this gap narrows considerably at later ages, but it does not close completely.

To help reduce this gap, we suggest that boys might benefit from extra training in language and study skills during the early school years. The most efficient way to improve overall performance would be to provide such help evenhandedly to all children who need it, a group that would include more boys than girls.

Curiously, in the face of this female progress, the famous male advantage on the mathematics section of the SAT has not narrowed at all (d' = 0.4, thirty-five points as of 2009). Despite their lower average SAT scores, however, women get better grades than men in college math classes. One study found that male freshmen get the same grades in college math classes as women whose math SAT scores are thirty-five points lower. Indeed, nearly all the standardized tests required for college admission underpredict the future grades of women. This poor prediction may be due to better study habits among women (giving them higher grades for the same aptitude), or it may be due to gender bias in the SAT (giving women lower scores for the same aptitude). Either way, the lower grades of college men relative to women can be attributed in part to the use of standardized tests for admission decisions.

Another well-known group of sex differences falls in the realm of emotional behavior. These differences are not as large as most people believe. Effect sizes range from small to medium. These differences do not predict individual behavior very well, but some of them are noticeable at the group level. Girls are more likely than boys to express fear and to cry, but in both sexes the physiological responses to distress are similar. Many differences are so small that they are drowned out by individual variability within each sex. One example is the idea that boys make moral decisions based on justice (d' = 0.19), while girls make moral decisions based on relationships (d' = 0.28). Similarly, girls are only slightly better than boys at identifying emotions in other people's faces (d' = 0.19 in childhood and adolescence). Boys are only slightly more likely than girls to take risks at all ages (d' = 0.13), with a larger effect between ages ten and thirteen, and between ages eighteen and twenty-one (d' around 0.25). The gender gap in risk taking seems to be closing over time, as it was smaller in later studies (1980s and 1990s) than in earlier studies (1960s and 1970s).

Kids return to mixed-sex socializing as teenagers. The hormones of puberty usher in a medium-sized sex difference (d' = 0.53): 70 percent of teenage boys masturbate more often than the average teenage girl, a pattern that continues into adulthood. The size of this difference has declined from a d' of 0.96 over the past two decades, suggesting a strong cultural influence. Sex differences in self-esteem peak in adolescence as well, with teenage girls showing lower self-esteem than teenage boys (d' of 0.33, another small difference).

One area where girls could clearly use extra support is body image. Especially as teenagers, girls experience much more dissatisfaction than boys with their bodies, which is a risk factor for eating disorders and depression. The size of this difference increased from a small d' of 0.27 in the 1970s to a moderate d' of 0.58 in the 1990s, perhaps because of the progressively thinner standards of female beauty across recent decades.

Even if you're concerned about your daughter's weight, criticizing her body is likely to be counterproductive. In one **longitudinal study** (in which the same people are followed over a period of time), teenage girls and boys who reported being teased about their weight by family members were much more likely than average to have developed an eating disorder or to have become overweight five years later. Similarly, in another longitudinal study, repeated dieting in fourteen-year-old girls (many of whom were not overweight when the study began) increased their risk of becoming overweight a year later by almost a factor of five.

One interesting thing about all of these sex differences is that the size of the difference does not predict how malleable it is. Though initial preferences can be modified by environmental influences, they often do launch boys and girls onto paths that can lead them to have different experiences for much of childhood. To sum it up, your child probably has some initial inclination toward fixing cars, taking care of babies, or whatever, but there are many opportunities to broaden your child's horizons by introducing him or her to new interests.

Chapter 9

ADOLESCENCE: IT'S NOT JUST ABOUT SEX

AGES: TWELVE YEARS TO TWENTY YEARS

You might dread your child's adolescence, fearing a tumultuous period dominated by hormones and erratic behavior. But the truth is far more complex and includes many other changes, which are overwhelmingly for the good.

Although key steps in sexual maturation do occur during this time, a host of changes unrelated to sex also take place before and after puberty. More than anything else, the adolescent brain is highly dynamic. During adolescence, which begins with the onset of puberty, usually between ages eleven and thirteen, and continues until twenty and sometimes beyond, children make major moves toward living on their own. They explore new interests, organize their own behavior, and pursue serious relationships outside the family. They revel in (or feel awkward about) their bodies' new capabilities. Most people recall their teen years as a time of near limitless possibility, of idealism, and of innumerable options. Friends of ours often find their early teenage daughter up late studying Spanish verbs, working on an intricate and beautiful drawing, looking up song lyrics, doing a conditioning regimen for her circus aerials, or simply reading or thinking. Whew.

Adolescence is also a time of risk. Just as developmental events before and after birth can lead to disorders such as autism, other problems are likely to become apparent during adolescence. Depression, bipolar disorder, drug addiction, and schizophrenia become increasingly prevalent at this time. In addition, adolescents are prone to take risks because their sensation-seeking impulses become strong when self-regulation is not yet fully mature.

To superficial appearances, the brain appears to be nearly finished as children enter adolescence. By late childhood, the brain has reached 95 percent of its adult volume. Individual components are within 10 percent of adult size (some larger, others smaller). Behind this apparent maturity, though, some large changes are stirring.

The adolescent brain undergoes considerable reorganization as synapses are pruned away, continuing the process that began in childhood. The brain contains its maximum number of synapses (the connections between neurons) well before puberty, in people as well as other primates (see chapter 5). Studies of brain glucose consumption in children, as well as detailed counts of synapses, show that by early adolescence, the human neocortex has reached adult synapse numbers and uses about one fourth less energy than it did in early childhood. Even so, synapse elimination is far from complete. Indeed, measurements from rhesus monkeys show that their brains lose as many as thirty thousand synapses per second during adolescence. In our larger brains, the number is probably higher.

Before getting into what your adolescent's brain is up to, let's get technical for a bit. To explain how and why your child's behavior is changing, we need to give you some details at the level of cells and connections that will provide essential context.

As you might expect, the changes in synapse number are accompanied by visible changes in the gray matter of the neocortex, where neurons, dendrites, and synapses are found. The general pattern is for gray matter to reach a peak thickness and then decline by 5–10 percent. In this way, the brain's circuits are shaped and refined before adulthood—while the brain's owner is acquiring new abilities to contribute more actively to the process.

These maturational changes happen at different times in different parts of the brain. Overall, the gray matter, containing all the neurons and synapses as well as a great deal of wiring, reaches peak volume by age nine to eleven. During this time, the **white matter** is still growing. In the neocortex, the first areas to reach peak thickness are the most extremely frontal and occipital (farthest back) regions. The areas in between are then filled in, starting from the back and moving forward. The temporal cortex reaches maximum thickness around age fourteen, followed by most of the frontal cortex. Finally, the white matter, made

of myelinated axons that carry long-distance information, bulks up, especially connections between frontal and temporal cortex.

One sign that adolescent brains are becoming more efficient is that activity is better coordinated between distant brain areas. This improvement is seen in signals varying together (coherency) and traveling over distances more quickly. White matter is only 85 percent of adult size and continues to grow even into the forties. As white matter grows, axonal fibers are likely to be widening, and fatter axons transmit signals at higher speeds. Because white-matter axons mediate communication between distant brain regions, this change is likely to have strong functional implications—though at present we don't know what they are.

The tempo of developmental change varies from child to child. In a study of children whose brains were imaged repeatedly as they passed through late childhood and adolescence, children of higher intelligence had gray matter thickness that rose to peak more steeply—and declined more quickly as well. This result suggests the possibility that a key to intelligence is not brain size but capacity for change, though these differences are too variable for evaluating individuals. Indeed, increases and decreases in gray matter thickness also appear in childhood-onset schizophrenia and ADHD, so these structural changes may reflect a variety of underlying processes in different children.

To superficial appearances, the adolescent brain appears to be nearly finished. In fact, it is undergoing considerable reorganization.

What do all these brain changes mean for your adolescent's thought—or lack thereof, as the case may be? The relatively late maturation of the frontal cortex has received a lot of media attention recently as a way of explaining adolescent impulsivity. Even one car insurance advertisement points out that this brain region is not done growing. This area participates in **executive function** tasks, such as self-control, planning, and resisting temptation (see chapter 13). It becomes more active with age, an exception to the general trend of decreasing activity. In anterior and superior regions of the frontal cortex, activity rises from ages twelve to thirty. In combination with the earlier maturation of **subcortical** areas participating in emotion and reward, adolescence is a time when the balance between impulse and restraint may be quite different from either childhood or adulthood.

Adolescents seek novel experiences more often and weigh positive and negative outcomes differently from adults. Such judgments can be probed using the Iowa Gambling Task, a game in which people can pick cards from several decks to win play money. Without the player's knowledge, some decks are stacked, leading to more losses overall, but large occasional gains. In a version of the game in which participants can play or pass, adolescents learn to prefer winning decks but are less prone to avoid losing decks. Only in their late teens do players show full avoidance of bad outcomes. In this game, then, adolescents make decisions that recognize the possibility of a lucky win but give little weight to losing.

This laboratory finding is reminiscent of the real-life observation that teenagers tend to underestimate the consequences of their actions. This tendency, noted since ancient Roman times, is seen in areas as diverse as unprotected sex, experimenting with drugs, and impulsive speech. Even though adolescents are physically healthy, this risk taking makes the mortality rate of this life phase high. Sam is fortunate to have survived his own youth, during which he habitually returned very late at night from social outings—the reward. Once he got into a bad car crash—a risk of staying up to the point of drowsiness unforeseen by his adolescent brain, which was focused on the short term.

These forms of impulsivity come at a time when white-matter connections between the frontal cortex and other parts of the brain that handle reward and emotion are not yet complete. Teenaged laboratory subjects are more likely to take a risk in games where there is a possible reward (money or the display of a happy face). When placed in an **fMRI** scanner, the teenagers showed more activity than adults do in the **ventral striatum**, a region that can signal anticipation of a reward. Another late-maturing participant in impulse control, the **orbitofrontal cortex**, appears to orchestrate the connection between emotion and good judgment.

In general, decisions are often informed by the brain's evaluation of whether an outcome is desirable or undesirable. Such a decision carries some emotional weight—even when it's as simple as picking an outfit to wear. People with damage to their orbitofrontal cortex are unable to sensibly manage their lives, making bad investments and unsuitable life choices. One patient (known by his initials EVR) had a benign tumor pushing on his orbitofrontal cortex. He lost his job, left his wife, married a prostitute, and divorced again in a matter of months. Removal of the tumor reduced these unadaptive behaviors.

Adolescent changes in mood, aggressiveness, and social behavior are driven by other aspects of brain development. These changes may be linked to increases in size and activity of the amygdala, a part of the forebrain that processes strong emotions, both positive and negative. Even puberty itself is ultimately driven by the brain, because the hypothalamus, a grape-sized structure that sits under the brain just in front of the brainstem, secretes **gonadotropin-releasing hormone** as the first step in a chain reaction that ultimately leads to the release of estrogens and testosterone to drive sexual maturation. Together, these hormones

MYTH: ADOLESCENTS HAVE A LONGER DAY-NIGHT CYCLE

The eight-year-old who got up early every morning has turned into a sluggish teenager. Although his body is in front of you, his brain is at least one time zone to the west. Everyone else is getting up, but he still wants to sleep—a kind of Adolescent Savings Time. What is going on?

Our brain's circadian rhythm sets the times that we want to wake and sleep (see chapter 7). Individuals vary, so that *larks* have peaks and troughs earlier in the day than *night owls*. Adolescence is accompanied by a shift toward evening wakefulness—and not just in people. At puberty, a shift of one to four hours has also been seen in monkeys and a variety of rodents.

One popular view is that adolescents have a longer day-night cycle. This impression is false; if you take away normal light-dark signals or suddenly shift the signals, a teenager's internal clock will react the same way as everyone else's. But a real difference in adolescent circadian rhythms is a decrease in melatonin levels, as well as a shift in the time when melatonin rises and falls. Melatonin helps trigger the onset of sleep. When puberty hits, nocturnal melatonin levels decline sharply, continuing a general decreasing trend that started back in infancy. So it's possible that adolescents are simply experiencing smaller and later sleep signals than they did in previous years, leading to a delayed bedtime.

With puberty also comes new social pressures. Even though they need only a little less sleep than children, adolescents are expected (or want) to adopt adultlike wake and sleep times. Their schools convene earlier in the morning. At the end of the day, there is homework, after-school activities, and spending time with friends. Intellectually and socially, their world is exploding. Even after bedtime, communications such as text messaging provide a continuing source of stimulation—and sleeplessness.

The net result is the need to catch up on lost sleep. In one study, researchers surveyed sleep habits in Swiss, German, and Austrian girls for up to nine years after their first menstrual period. The girls slept almost two hours longer per day on weekends than on weekdays, compared to less than an hour of catch-up in younger children and adults. Sleep debt has serious consequences, including reduced mental performance, depressed mood, impaired health, and weight gain.

One name for this adolescent tendency, *social jetlag*, suggests that they might be able to use some tricks of long-distance travelers. Here are a few:

1. Opening the blinds in the morning will activate the melanopsin pathway in your retinas. At this time of the circadian clock, exposure to light creates a tendency to get up a little earlier the next day.

2. Evening light leads to a later bedtime the next day. Combine that with a natural tendency to stay up, and it's a recipe for continued night-owl behavior. So even if sleep isn't coming easily, turn down the lights. And turn off that cell phone!

3. Exercise leads to secretion of melatonin by the pineal gland. An evening soccer game or run might be just the thing to start a brain on the road to sleep.

---

powerfully reorganize the brain. Many of the brain changes we have described may be organized and shaped by hormone signaling.

Although sex and stress hormones rise during late childhood and adolescence, in most cases researchers have found little evidence for a direct effect of hormones on typical adolescent behavior. Hormones are a key component for organizing the neural circuitry, but by itself testosterone is not very predictive of risk taking. The combination of a poor parent-child relationship with high testosterone has somewhat more predictive power. In adolescence, a good relationship forged in your child's early years can pay off. This principle extends to siblings too: better relationships with brothers and sisters improve adjustment during adolescence.

A degree of impulsivity and aggression is probably unavoidable in life, but in some cultures, adolescent urges play a positive role. For example, among immigrants in big-city Chinatowns, aggression by male adolescents toward potentially violent intruders can protect the community from harm. Among the Mbuti, a hunter-gatherer group in the Congo, adolescents act on behalf of the group to punish deviations in adult behavior with mockery and even vandalism.

One hallmark of adolescent behavior in people and other mammals is an increase in what behavioral scientists call *approach*, the seeking of new social contacts and situations. Combined with other changes, this tendency can lead to the making of new friends—and also, sometimes, rebellion against older family members. Some conflict is typical, though extreme emotional turmoil in relationships with parents is experienced by only about one in ten adolescents. This happens in other species as well. For instance, adolescent rats sometimes attack their parents.

The tendency to seek novelty is likely to be driven by the brain's reward systems. **Dopamine** is a neurotransmitter involved in initiating action and movement and in signaling rewarding events. In brain scanning data, the orbitofrontal cortex and other regions that receive dopamine-secreting inputs are still maturing during the teen years. The *serotonergic system*—involved in sensation, movement, and mood—is also changing in the adolescent years. Awkwardness and moodiness might be linked to this change.

Another change in the brains of adolescents is a proliferation of receptors for the signaling chemical *oxytocin.* (Oxytocin is a **neuropeptide**—that is, a peptide used as a neurotransmitter.) Neuroscientists found that oxytocin mediates a wide variety of bonding behaviors. In people, oxytocin is secreted during feelings of romantic and parental love. Both mothers and fathers of an infant or a small child have more oxytocin; the higher their oxytocin, the more they touch, play with, and otherwise socially engage with their child—and each other. Indeed, these signals sometimes get crossed, so that a new mother having a loving thought toward her partner might feel her milk drop. Romeo and Juliet would also have felt a newly strengthened oxytocin signal.

Adolescence is a time when the interplay between brain and environment takes on new complexity. Early adolescent brain changes increase a child's appetite for stimulation and social contact, while self-regulatory systems continue to mature through late adolescence. In modern society, adolescence is viewed in terms of the delay between sexual maturation and true independence. Indeed, sexual, physical, and intellectual maturation are spread out over a decade or more, providing many opportunities for growth and change. What an adolescent does with this biologically defined period of transition depends on his or her culture—and the choices that come along the way. Around the world, how and when people enter society during this process varies, ranging from child workers to continuing students with children of their own. In all cases the brain has found ways to adapt to local circumstances—a testament to its flexibility.

PART THREE

START MAKING SENSE

Chapter 10

LEARNING TO SEE

AGES: BIRTH TO FIVE YEARS

As you haul your kids from music lessons to swim team practice, have you ever paused to give thanks that you don't have to take them to vision class? Learning to see is a complicated process involving the coordinated development of dozens of brain areas. It does depend on experience: children's eye problems can affect their ability to see. However, most parents get an almost completely free ride when it comes to their children's eyesight.

This doesn't mean your baby arrives ready to see. Far from it. An adult who could see as well as a newborn would be legally blind, with 20/600 vision. Babies' visual acuity starts out forty times worse than adults' and doesn't become equal until four to six years of age.

Like other aspects of your child's brain development, this maturation comes about as an interplay between genes and experience. In this case, the necessary experience is visual and is widely available to any baby who can see normally. Most baby books don't talk much about this type of development (which scientists call *experience expectant*), perhaps because it doesn't require or reward much effort from parents. But as we outlined at the beginning of this book, such robust (self-managing) processes are more the rule than the exception in early life. Experience-expectant development is one of the key reasons that most kids end up well adapted to their environment.

Though vision feels seamless, your brain actually constructs its image of the world from the neural activity in dozens of interconnected regions that specialize in particular aspects of seeing. These regions are organized into two main pathways. The "where" pathway, which develops earlier, consists of the cortical areas that process motion and space. The "what" pathway is made up of areas that evaluate the properties of objects, including their shape, color, and patterning. Both pathways obtain information from a chain of connections that start at the retina, pass through the thalamus, and go to the primary and secondary visual areas of the cortex. From there, the two pathways diverge, involving different parts of the cortex, but with plenty of crosstalk between them.

All these cortical areas are immature at birth, making the vision of newborn babies quite poor. Babies don't see what we see. Newborns mostly rely on subcortical pathways, from the retina to the **superior colliculus**, a midbrain region that controls visual-motor reflexes (like flinching away from an incoming object) and certain types of eye movements.

When the visual cortex starts to mature in the second month of life, it takes control from the subcortical pathways. This transition often does not go smoothly. At this age, many babies show *obligatory looking*, the inability to pull their gaze away from something that has caught their attention, sometimes for as long as half an hour. This difficulty is caused by the visual cortex inhibiting subcortical eye movement commands. Young babies track movement with jerky eye movements called *saccades* until two or three months of age, when cortical maturation allows them to smoothly follow a moving object with their eyes. In the first three months, babies also have difficulty focusing their eyes on faraway scenes, so they look at what is nearby (roughly seven to thirty inches away), which includes their own bodies and their parents' faces.

The champion of the infant visual system is motion, which develops early and effectively. Babies can detect a flickering stimulus in a single location almost as well as adults as early as four weeks of age, and the flicker frequencies that they can detect become adultlike by two months. To determine motion direction, it's necessary to associate time-based changes coming from different locations in space, a capacity that appears around seven weeks. By twenty weeks, babies can discriminate different speeds of motion. Perception of large-scale motion patterns, like raindrops seen through the windshield of a moving car, improves rapidly between three and five months and then continues to develop slowly through middle childhood. This aspect of motion processing, the most vulnerable to disruption, is impaired in some developmental disorders, including dyslexia and autism.

PRACTICAL TIP: **OUTDOOR PLAY IMPROVES VISION**

The stereotype of the nerdy guy with glasses has some basis in fact. Myopia (or nearsightedness) has the curious quality of being both inherited (with a heritability of about 80 percent) and strongly influenced by the environment. How this happens is a lesson on the complex ways that genes and environment can interact.

Myopia occurs when the lens of the eye focuses the visual image in front of the retina, causing faraway objects to look blurry. The incidence of myopia varies tremendously across populations, from 2–5 percent among Solomon Islanders in the 1960s to 90–95 percent among modern Chinese students in Singapore. The rate of myopia has increased considerably over the past few decades in many countries. In Israel, 20 percent of young adults were myopic in 1990, increasing to 28 percent in 2002. Similarly, in the U.S., myopia rates went up from 25 percent in the early 1970s to 42 percent in the early 2000s. These changes have been happening so fast that the explanation couldn't possibly be strictly genetic—some external factors must be involved.

As your child's eyes grow, the distance between the pupil and the retina needs to be matched to the focusing power of the lens to keep the image on the retina sharp. If this distance is wrong, either myopia or farsightedness results. Based on experiments with animals, we know that visual experience guides this process.

Children who spend more time outdoors are less likely to become myopic. One study compared six- and seven-year-old children of Chinese ethnicity living in Sydney, Australia, with those living in Singapore. The rate of myopia was more than eight times lower in Sydney (3.3 percent) than in Singapore (29.1 percent), despite similar rates of parental myopia (about 70 percent in at least one parent). Children in Sydney spent fourteen hours per week outside, on average, compared with three hours per week for children in Singapore.

It does not seem to matter exactly what the children do while they are outside. A U.S. study found that two hours per day of outdoor activity reduces the risk of myopia by about a factor of four compared with less than one hour per day. Playing indoor sports has no effect on visual development. Outdoor activity has a stronger protective effect for children with two myopic parents than for children with no myopic parents, suggesting that myopia-related genes may modify children's sensitivity to environmental influences (see chapter 4).

Researchers are not sure why being outdoors protects children against developing myopia, but one possible explanation is that bright outdoor light is more effective than dim indoor light at driving the development of correct pupil-retina distance. Since our brains evolved under conditions in which every child spent many hours outside every day, it makes sense that our eyes may develop in a way that takes advantage of that common experience. Our current lifestyles may well lead to other unexpected consequences of this sort, as our brains are forced to adapt to a world that's very different from the one in which their genes originated (see *Speculation: Modern life is changing our brains*, p. 88).

The vision of infants is partly limited by the maturity of rods and cones, which translate light into neural signals in the retina. Cones, which provide sensitivity to color, mature rapidly. Though color vision is almost absent in newborns, four-month-old babies can see color as well as adults. Rods, which do not transmit color but detect photons in low light (which, incidentally, is why you can't make out colors in the dark), mature by six months. Newborns can see better in their peripheral vision than in the center of their gaze, both because the cones in the peripheral retina are more mature and because cells in this part of the retina project more strongly to subcortical visual areas.

Visual acuity is easy to test because babies prefer to look at patterns. Researchers can tell whether babies can distinguish a pattern from solid gray just by whether they preferentially look at the patterned object. At three months of age, babies are still fifty times less sensitive to contrast than adults, meaning that infants find it very difficult to distinguish different shades of gray. It's like they're seeing the world through dense fog (see figure below). These limitations explain

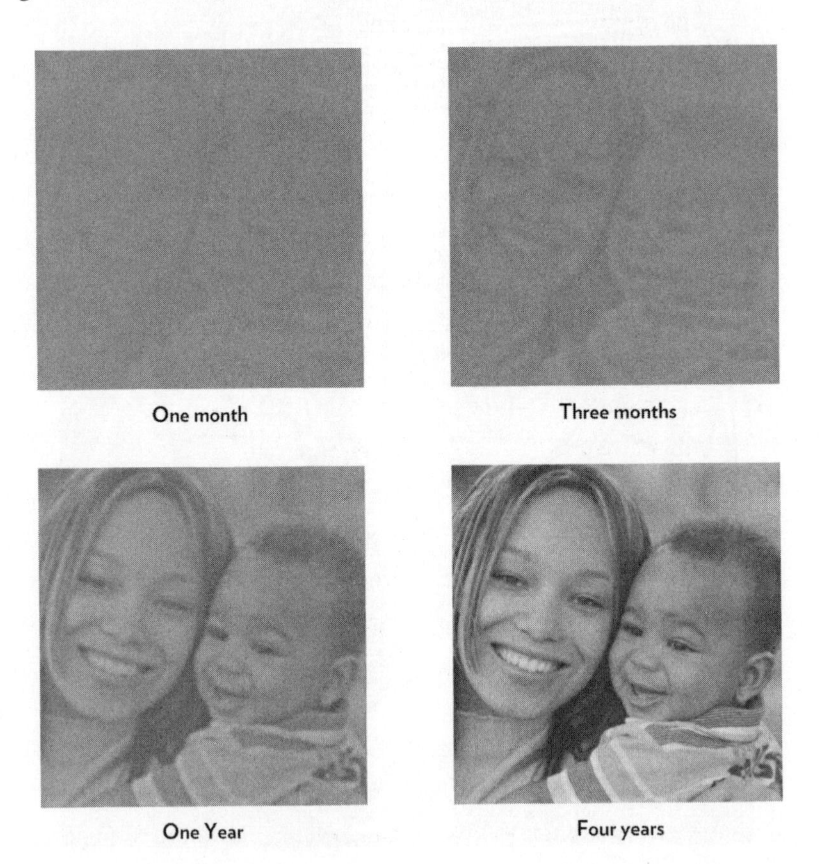

why the most popular toys for young infants have bold black-and-white patterns.

For depth perception, both eyes need to work together. It's very difficult to thread a needle, for instance, with one eye closed. The two eyes don't see exactly the same part of the visual world, and the differences between them depend on head size. The brain uses visual experience to sort this out as children grow. Depth perception is almost nonexistent in newborns. The ability to use binocular cues develops abruptly, often in the fourth month of life.

An adult who could see as well as a newborn would be legally blind, with 20/600 vision.

From birth, babies are attracted to faces. It may be no coincidence that babies can focus best on objects about eight inches away, which is approximately the distance between the baby's eyes and the parent's face during feeding. Very young babies, though, are working from an approximate model of what a face looks like, as they will look at almost any round thing that has two "eyes" and a "mouth" in the right place. (This is not very surprising if you consider how poorly they see real faces.) By four or five months, their preferences are more realistic, and babies have begun to process faces differently from other objects. This change probably reflects maturation of the **fusiform face area**, a region in the temporal cortex specialized for face processing. This brain specialization enables ordinary adults to beat the world's best computer programs in detecting subtle differences between faces. The fusiform face area appears to already be preferentially activated by faces in two-month-old infants.

Development of many visual functions requires experience during a sensitive period (see chapter 5). Early in cortical development, chemical cues direct axons from each visual area to innervate its appropriate target areas, where they form many more synapses than will be needed in the adult brain. Neural activity patterns then control axon retraction and synapse elimination, which fine-tune the connections so that the correct neurons talk to each other. In the primary visual cortex, for example, synapse number is greatest at eight months and then declines through middle childhood. Because different brain areas develop at different ages, the effects of visual deprivation vary with timing.

Children whose vision is impaired by cataracts provide information about the

SPECULATION: MODERN LIFE IS CHANGING OUR BRAINS

For millennia, children could reliably expect to have certain experiences. Infants would hear their parents and other adults talking. Babies would see objects, some of them colored, some of them moving. Food would be obtained from nearby land. The sun would bring light when it rose and leave darkness behind when it set. Our brains evolved to make the most of these situations.

But times have changed. Since the invention of agriculture, and especially since industrialization, the environment has changed substantially and in many cases has come under our control, making some of these realities a lot less reliable. What happens when experiences necessary to our development are hard to find? Artificial light is much less bright than sunlight and seems to interfere with the normal matching of lens power to eye size by experience (see *Practical tip: Outdoor play improves vision*, p. 84). Grocery stores are full of processed food, which lacks fiber, nutrients, and variety compared to our ancestral diet. Our brains have evolved to seek out sugar and fat because such foods were rare treats during our evolution, but now they are readily available. These dietary changes may contribute to the rise in obesity and some types of cancer.

These examples illustrate a fundamental conceptual problem with trying to separate the effects of genes from the effects of the environment: the two influences are inextricably linked (see chapter 4). Evolution has selected for genes that produced an adaptive outcome in our ancestral environment, but these genes may not interact as effectively with our current environment.

That doesn't mean that there's anything wrong with the modern world (we like our computers and antibiotics, thank you very much), nor that there's anything wrong with our genes; it's just that they don't play nicely together in some cases. For instance, type 2 diabetes, which is linked to a variety of lifestyle risk factors, is also highly heritable. This may seem less confusing if you think of genes and environment as having a conversation about how growth should proceed. In this framework, particular genes and certain environmental conditions can easily interact to produce an unfavorable outcome that would not have resulted from variations in either the genes or the environmental conditions alone.

need for visual experience in human development. Babies who have cataracts from birth retain the poor acuity of newborns until their eye function is surgically restored, even as late as nine months of age. After that, with experience their acuity improves, but it does not catch up fully. Deprivation for the first three to eight months leads to acuity more than three times worse than normal at five years of age. Children who develop cataracts later on, starting between four months and ten years and lasting for two to three months on average, also end up with permanent deficits in acuity but are not as impaired as babies whose cataracts were present from birth. Global motion perception is affected by cataracts only in the first three months of life.

Visual experience also influences the development of face-recognition expertise starting in infancy. Six-month-olds are as good at distinguishing individual monkeys as individual people, but by nine months, babies become better at distinguishing people and lose the ability to discriminate among monkey faces. Starting between six and nine months of age, babies also find it slightly easier to distinguish faces within their own racial group than within other racial groups, probably because most babies have more visual experience with their own racial group than with others. This process, which is reminiscent of phoneme learning (see chapter 6), probably involves the sculpting of synaptic connections by experience to tune perception to the characteristics of the local environment.

Because our abilities build on one another, sensory deprivation during a sensitive period early in life can initiate a cascade of problems later on. This also means that aspects of vision that develop later are more sensitive to disruption than those that develop earlier. For example, babies cannot distinguish fine stripes from solid gray until their second birthday. Even so, babies who have cataracts from birth to six months of age never develop this ability, apparently because of residual damage to their primary visual cortex.

These findings suggest that parents should be particularly careful to protect their children from sensory deprivation early in life. A variety of problems, including cataracts, amblyopia, or strabismus, can prevent babies and toddlers from getting the experience that they need to help the visual system develop correctly. Strabismus occurs when the two eyes do not point in the same direction, which interferes with the development of binocular vision. Amblyopia occurs when one eye sees substantially less well than the other eye, though both are apparently healthy. It can be caused by strabismus or by near- or farsightedness that occurs in only one eye. While strabismus or cataracts can be diagnosed by parents or doctors based on the appearance of the child's eyes, amblyopia can only be identified by trained professionals.

Routine well-baby exams should catch most problems of this sort, but if your child is diagnosed with a sensory deficit (or you suspect that one exists), getting it fixed as soon as possible will minimize the possibility of lasting damage to your child's brain. Amblyopia can be treated by several methods. Corrective glasses should be the first step, as this approach solves the problem in a quarter to a third of children and improves the amblyopia in others. The next step is to put a patch over the strong eye to force the weak eye to work harder. For all early visual difficulties, the most important point is to act quickly, before serious damage occurs.

Fortunately, problems in visual development are the exception. It's amazing when you consider how complicated this process is, but in most cases, parents really can simply sit back and watch their children's new abilities grow.

Chapter 11

CONNECT WITH YOUR BABY THROUGH HEARING AND TOUCH

AGES: THIRD TRIMESTER TO TWO YEARS

We have big brains. This is one of the things that distinguishes humans from other primates, and it has many consequences, some of them unexpected. One is that, because women's hips are only so big, our babies have to be born before their brains have grown to full size.

As a result, the brains of infants are very immature. Newborn infants can't roll over by themselves, and they can barely see, as we learned in the previous chapter. Some parts of their brains are relatively well developed at birth, though, including those devoted to the senses of hearing and touch, which provide the best ways to connect with your new baby.

Hearing begins when the ear receives a sound, a set of pressure waves that move through the air the way a splash ripples across a pond. The time between arriving waves (frequency) determines pitch, and their height determines sound intensity. The outer ear transmits these waves to the *cochlea*, which contains sound-sensing *hair cells* arranged along a long, coiled membrane (see figure, p. 92). Sound pressure moves the fluid in the ear, which makes the membrane vibrate differently depending on the sound's frequencies. This vibration moves a bundle of fine fibers that stick up from the top of the cell (hence the name hair cell), transforming the vibration into an electrical signal that can be understood by other neurons.

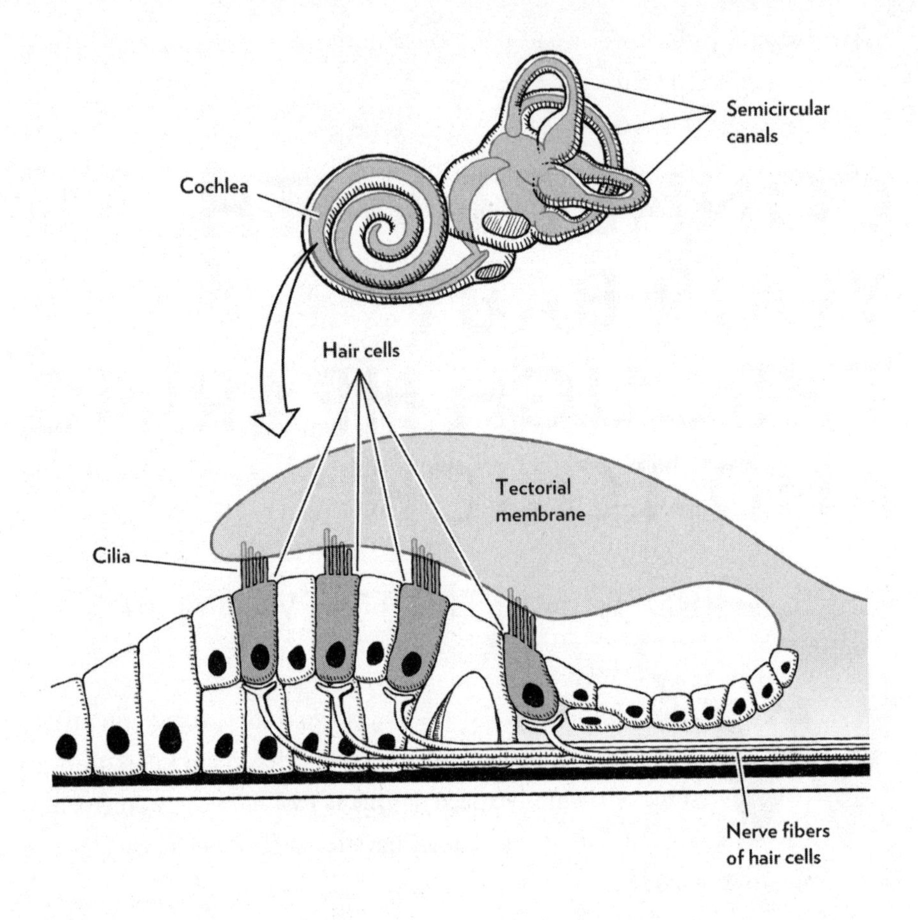

Hair cells at the base of the cochlear membrane sense the highest frequencies. As you move around the coil toward the other end, hair cells sense lower-frequency sounds. This organization forms a map of sound frequency, which is maintained in many of the brain areas that receive information that first passes through the cochlea. As in visual development, experience is important for fine-tuning connections in auditory brain regions, but the appropriate experience is easily available to any child who can hear.

A related group of organs in the inner ear, the *semicircular canals* (see figure), is responsible for sensing the head's acceleration. This is called the *vestibular system.* These canals also contain hair cells arranged in a circle, where they are stimulated by calcium crystals—essentially little rocks that roll around inside the ear and settle on top of particular hair cells depending on which way is up. Imagine the beads inside a baby's rattle; if you could feel the position of the beads, you'd be able to work out which way it was pointing. This is what the brain is doing with information from the semicircular canals.

The vestibular system matures early, in the second trimester. It is vulnerable to many of the same factors that can disrupt hearing, including prenatal infections—particularly cytomegalovirus, a disease that accounts for 12 percent of congenital deafness—low birth weight, and bacterial meningitis in infancy. Disorders of the vestibular system can lead to developmental delay in motor function, as it is very difficult to learn to walk without reliable balance.

Babies can hear before they're born, starting around the beginning of the third trimester. At this stage, they can hear only loud sounds at medium to low pitches—like a car horn or a truck rumble—which is convenient because those are the sounds that most easily reach the baby through the insulation of the mother's belly. The mother's voice also reaches the baby's ears strongly because it is carried within her body. With time, the auditory system gradually becomes sensitive to quieter noises and higher pitches, a process that continues after birth.

Auditory learning is already occurring during gestation. By the time they're born, babies prefer their own mother's voice to a stranger's voice. Most newborns find the sounds that they've heard in utero to be soothing—anything from the theme song of Mom's favorite soap opera to her heartbeat. They also prefer the sound of her language to a foreign language, probably because its cadence is more familiar to them.

Living near a loud highway could damage your child's hearing as much as setting off firecrackers in her bedroom.

At birth, babies are still less sensitive than adults to quiet or high-pitched sounds. A normal conversation sounds to a baby about as loud as a whisper sounds to you. By six months, frequency sensitivity is fully mature, allowing babies to hear high-pitched sounds. However, at this age loudness thresholds are only halfway to adult levels. Children's ability to solve the *cocktail party problem* (discriminating a voice from background noise or from competing voices), a challenge that calls upon all of this information, continues to improve until age ten.

At all ages, children hear high frequencies better than most adults, whose hearing has been damaged by too many loud sounds (see *Practical tip: Protect your child from noise, starting before birth*, p. 96). Some teenagers take advantage of this situation by using high-pitched "mosquito ringtones" to prevent teachers and parents from hearing the ringing of cell phones in situations when they're forbidden. Adults have turned the tables by using loud high-pitched sounds to prevent young people from loitering outside businesses or in parks (though we do not support this approach because it may harm the hearing of infants and toddlers whose parents are unaware of the noise).

Given the importance of hearing in a baby's life, it is crucial to diagnose deafness in young children as early as possible. Nearly all babies are tested in the hospital at birth, but among those who are not, the average age of diagnosis is fourteen months, by which time auditory and language development are already delayed. Hearing loss can also develop in older babies due to infections or genetic problems with development. Parents should get their babies' hearing tested if they don't respond to sudden loud sounds or the voices of people who are out of their view.

Deafness resulting from damage to the cochlea can be treated with a cochlear implant, a device that transmits sound information directly to the auditory nerve. (If the deafness is due to damage in the brain rather than the ear, it usually cannot be corrected.) The signals sent by a cochlear implant are much less complex than the signals from a healthy cochlea, but the brain gradually learns to interpret the new signals correctly, especially if the implant is given in early childhood when the auditory system is best able to make adjustments. In general, cochlear implants can improve hearing at any age—but earlier is better.

The development of the sense of touch also depends on experience—in this case, provided by adult caretakers. Luckily, most people love to snuggle babies, so a lack of stimulation in this area is rare. Touch is critical for parent-infant bonding, which has an important influence on the baby's emotional and cognitive development. Early touch also influences adult stress responsiveness in many mammals, including people (see chapter 26).

The neural pathways that carry information from the skin develop early in gestation, before any other senses. Your baby's skin contains many different receptor types—specialized nerve endings that sense touch, vibration, pressure, skin tension, pain, or temperature (see figure opposite)—which are all in place by the end of the first trimester. Another set of receptors in muscles and joints provides

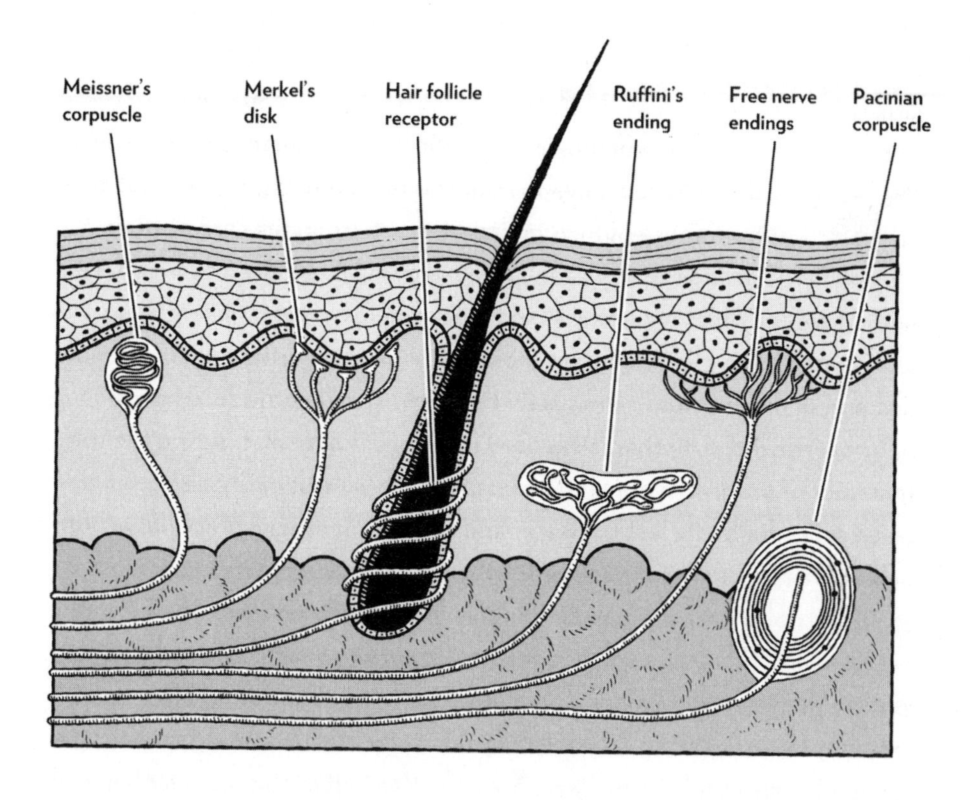

information about body positioning and muscle tension. In all cases, touch is translated into the spikes we've been telling you about, and sent along axons into the spinal cord or brainstem.

The brain pathways that process touch information are not fully formed until the fifth month of pregnancy. The brain knows which kind of sensor is activated, and where it is on the body, because each sensor has a "private line" that uses spikes to carry only one kind of information to the brain. (There is a separate brain area for painful sensations.) At birth, babies can identify touch and temperature sensations; they turn away from a cold object touching their cheek and toward a warm object. Because the axons that carry touch information are not yet myelinated, babies are eight times slower than adults in sensing most types of touch. Children's processing speed improves in the first year but does not reach adult levels until age six. The exception is pain, which is carried by unmyelinated axons even in adults and so is processed just as quickly by babies.

Some parts of your baby's body are more sensitive than others. As in other sensory systems, the cortical areas that process touch are arranged in a map, with

PRACTICAL TIP: **PROTECT YOUR CHILD FROM NOISE, STARTING BEFORE BIRTH**

Another problem in our modern environment that evolution did not prepare us to handle is noise exposure, the most common cause of hearing impairment. Its victims are getting younger every year. One in eight U.S. children between the ages of six and nineteen now has some hearing loss, and it's likely to worsen as they get older. Hearing loss is more common in boys than girls, probably because of differences in their activities. Noise can even induce hearing loss before birth, if the mother is exposed to chronic loud noise during the third trimester of pregnancy. In fact, babies' hearing is most easily damaged by noise exposure in the last trimester and first six months of life. Premature babies are especially vulnerable to noise-induced hearing loss.

Loud noise destroys hair cells in the cochlea, starting with the most vulnerable cells, those that transduce high-frequency sounds. Doctors cannot reverse this damage, and hearing aids do not restore sound levels. The earliest symptom of hearing loss is usually difficulty understanding speech when there is background noise. By then, as many as half of the cochlear hair cells are already dead. Hearing loss is particularly damaging in children because it can impair language learning and academic accomplishment.

Tinnitus, a constant ringing in the ears, is another potential consequence of noise exposure that can interfere with hearing. Noise is also a chronic stressor, which means tinnitus can interfere with child development in many ways (see chapters 26 and 30).

Hearing loss is caused by brief exposure to very loud noises, like firecrackers, or by chronic exposure to moderate noise levels, such as city traffic. (Living near a loud highway could damage your child's hearing as much as setting off firecrackers in her bedroom.) The most common risk factors for kids are rock concerts and portable music players like the iPod. These devices are typically played at 75 to 105 decibels, which is equivalent to the range between a loud conversation and a Harley-Davidson. Temporary hearing loss or tinnitus after listening to music is a warning sign; repeated exposure to such noise levels will cause permanent hearing loss.

The volume makes a big difference. With standard iPod ear buds, you can safely listen at 80 percent of maximum volume for ninety minutes per day, or at 70 percent of maximum for four and a half hours per day, but at full volume only for five minutes per day. Your kids should be especially careful when listening to music in a noisy environment, such as on an airplane or in the subway, which generates a temptation to turn the music up too loud. You can protect your children by downloading software that limits the music player's volume, or by investing in noise-canceling headphones (if they don't look too dorky for your child's sense of style). Your kids may not appreciate it now, but at least they'll be able to hear their own kids' complaints someday.

DID YOU KNOW? THE NEUROSCIENCE OF SNUGGLE

Different types of touch are transmitted to your child's brain by "private lines" that carry those particular types of information exclusively. Your child's brain (like your own) has a special pathway dedicated to the kind of touch that leads to emotional bonding. As we've said, the skin contains more than a dozen anatomical types of receptors specialized for detecting particular sensations, such as temperature, pressure, and pain. One in particular is tuned to the pleasurable skin sensations that are produced by light stroking. The axons that carry information about these sensations to the brain are unmyelinated, meaning that their responses are slow. In human recordings, the electrical activity of these axons in response to gentle stroking is proportional to how pleasant the person reports the touch to be.

Damage to these pathways as they run through the spinal cord impairs emotional responses to touch, without affecting the ability to identify an object by touch. In the brain, the "pleasantness pathway" brings touch information to a region of cortex called the *anterior insula*, rather than to the somatosensory cortex where most touch-responsive fibers send their signals. The anterior insula, which receives input from a vast variety of systems, seems to be involved in monitoring a range of internal states, from thirst to maternal love.

nearby parts of the body represented by nearby neurons. The proportions of these maps are based on the number of receptors in each part of the body, rather than its size, so that the part of the map that receives information from the face is larger than the area that receives information from the entire chest and legs. Your face is much more sensitive. In adults, the highest density of touch receptors is found on the fingertips, with the face a close second.

This map develops sequentially, beginning at the head region, which is why newborn babies actively explore objects with their mouths but not with their hands. Think about the surprising number of familiar objects whose flavor you can imagine—doorknobs, grass blades, and so on. Evidently you've been tasting things for a long time.

The ability to discriminate objects with the hands develops slowly, and the face remains more sensitive than the hands even at age five. As the maps in the somatosensory cortex mature, babies become able to localize touch more accurately and discriminate touches that are closer together on the skin. These maps are initially established by genetic mechanisms, but their maintenance depends on experience even in adulthood. For example, the cortical space devoted to an amputated limb is eventually taken over by inputs from adjacent areas of the body (the cause of phantom limb syndrome, in which amputees imagine they can feel sensations from a missing limb).

Babies who are not touched enough in early life become developmentally delayed, demonstrating one limit to the dandelion nature of brain growth. This problem happens most often in poorly organized institutional care, such as orphanages with inadequate staffing or intensive care units in which premature infants are isolated from human contact. Cuddling is more important than food to early bonding; in experiments, baby monkeys deprived of maternal contact spent most of their time with a soft surrogate made of terry cloth, ignoring a wire surrogate that provided milk except during brief visits to feed. In most homes, though, you're more likely to have trouble getting family members to stop playing with the baby at bedtime than you are to have a baby who's not getting enough snuggling.

Chapter 12

EAT DESSERT FIRST: FLAVOR PREFERENCES

AGES: SECOND TRIMESTER TO TWO YEARS

Unlike many American toddlers, Sam's daughter loved sushi. From a young age, she made it impossible for her parents to eat their raw fish in peace, without little orange fish eggs flying everywhere. Though we can't say for sure, we suspect that a trip she took before she was even born might be responsible for this odd turn of events.

Sam and his wife traveled to Japan in the second trimester of her pregnancy. Sam's wife adores sushi. As a physician, she knows that sushi is safe for babies in utero—and perhaps even beneficial for brain development (see *Practical tip: Eat fish during pregnancy*, p. 20). So she ate a lot of it on that trip and afterward. When we started reading studies showing that children's food preferences are influenced by what their mothers ate during pregnancy, we thought we might have found the connection.

Like vision and hearing, our basic ability to identify smells and tastes is built on sense organs and input pathways to the brain that develop largely on their own. Starting from our noses and tongues, the initial stages of input wiring connect with brain structures without much help. But much of the formation of preferences for smells and tastes depends on experience. As omnivores, people can eat such a wide range of foods that it would be difficult to genetically program innate food preferences that would apply to all possible environments. This may explain why food preferences are so idiosyncratic; for instance, many children in the U.S. like root beer, but in some parts of Europe it is considered intolerable.

The learning process starts surprisingly early, well before birth. Newborns have smell and taste systems that are well developed—and already know a thing or two. They can express an immediate preference for the flavor of milk and their mother's nipple, and they can even distinguish between the smell of their own mother and others. (Indeed, a mother can do the same, distinguishing a shirt worn by her baby from those of other newborns.)

At birth, amniotic fluid odors serve a bridging function, helping the infant coming out of the womb by soothing him and by helping to establish preferences for Mom (and her milk). During the first week of life, your baby's preference for the odor of amniotic fluid decreases, while a preference for natural breast odors increases. This preference for breast odor can even be helpful. A breast-fed infant attaches to the breast faster if it bears natural odor than if it has been washed.

So if nursing is an issue, you might try not washing. For similar reasons, if Mom wears perfume, switching perfumes may slightly confuse the baby at first.

In adults, smell and taste take a backseat to vision and hearing, which provide most of the information we use to navigate through the world. Infants, born with rather poor vision (see chapter 10), are more dependent on the chemical senses. They share this dependence with some of the oldest of animals. Over 800 million years ago, certain types of worm started to localize more refined chemical sensation in their front end—in other words, they developed noses. Even more primitive animals such as jellyfish are deaf, blind, and noseless, but can still detect the presence of noxious chemicals applied somewhere to the surface of their bodies.

The flavor sense consists of multiple components in the mouth and nose. What hits the tongue is taste and includes the fundamental components of sweet, salty, sour, bitter, and umami. (*Umami* is a Japanese word meaning "yummy." It is used to refer to the rich mouthfeel of cooked meat or mushrooms, caused by the presence of **glutamate**, a component of protein also found in tomatoes and many broths.) Combined with taste is the sense of smell, which comes in through the nose and up from the back of the mouth and is far more complex.

Smell conveys most of the nuances we experience in food. You can prove this to yourself using the "jellybean test." Take a bowl of jellybeans and eat one at

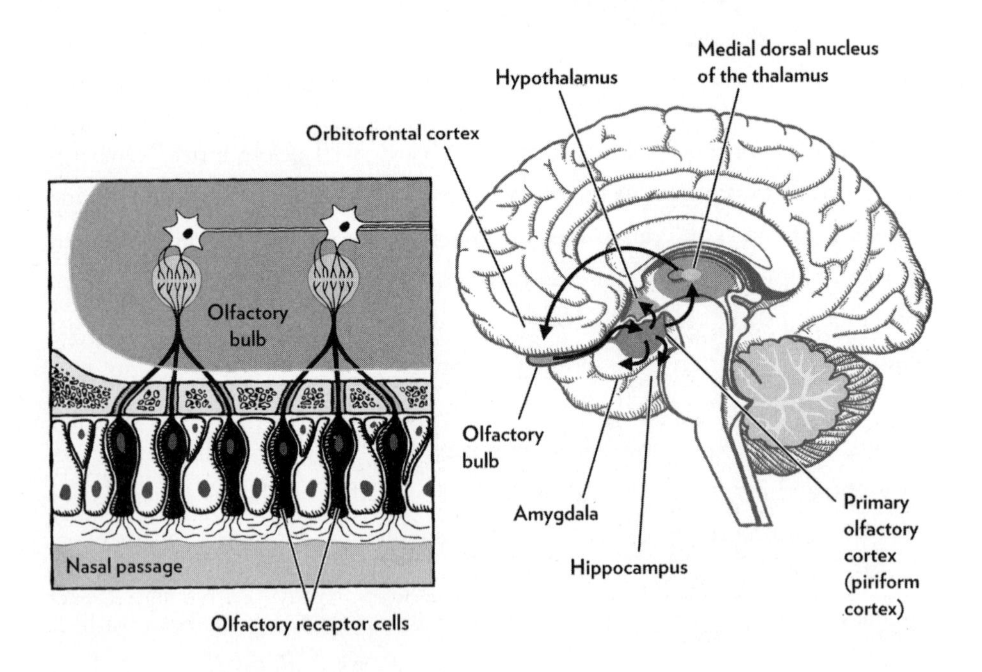

a time, first normally, then while holding your nose. While holding your nose, you will find that the candies taste much more similar to one another. The same is true for taking a bite from an apple compared with a potato, though you may not enjoy this test as much.

Smell signals are detected in the nose by a layer of receptor cells (see figure opposite) covered in sticky mucus that helps to catch odors. Each sensory cell expresses one type of odorant receptor protein. (All receptors are made of protein.) There are hundreds of distinct receptors, each with its own preference for a different set of odorant molecules. These sensory cells send their axons in a mad tangle, passing along the way through hundreds of perforations in a thin bone called the *cribriform plate*, which resembles a rooster's comb. The whole thing detangles to make a perfectly sorted landing pattern in the brain in the ***olfactory bulb***, an oblong structure that lies underneath the frontal part of the brain. By the time the axons arrive, each part of the olfactory bulb receives a "private line" corresponding to one, and only one, type of odor receptor.

So when can babies start to smell things? The olfactory epithelium and bulbs are present by the eleventh week of gestation, toward the end of the first trimester. Then, sometime during the second half of the second trimester (weeks sixteen to twenty-four), the nostrils open up, allowing amniotic fluid to reach the olfactory epithelium. In premature infants, the earliest reported responses to smell occur at the end of this period, at seven months of gestational age. This is possible not only because the nostrils are open but because neurons from the olfactory epithelium have sent axons into the brain, to the olfactory bulb. In turn, neurons in the olfactory bulb send connections to other places in the brain, including the amygdala, which is involved in generating emotional responses, and the *piriform cortex* (a region of the neocortex), which passes the information on to other brain regions (see figure opposite).

To an infant, most smells are initially neutral. Some responses are innate: within twelve hours of birth, infants make faces when exposed to the scent of rotting eggs. Likewise, sugary water triggers an automatic smile: more, please. Generally, though, smells take time to become associated with positive or negative reactions, emotions, and memories. These associations to the world's rich variety of food, drink, and smells are acquired postnatally. This process is guided by the brain's mechanisms for associating new smells with already-familiar signals and liked (or disliked) outcomes.

PRACTICAL TIP: GETTING YOUR CHILD TO EAT SPINACH

One in five American toddlers don't eat even one vegetable per day. Can we teach young children to like these bitter but healthy foods?

Most parents use social approaches, which often work: involving the child in preparing the food, or showing that parents or siblings like the food. Less appreciated are techniques based on the direct experience of flavors.

Just consuming a food multiple times is sufficient to reduce negative reactions. Infant taste is particularly plastic during the first few months. Babies fed a relatively bitter nonmilk (such as soy-based) formula are more tolerant of broccoli as children. And when asked to rank their favorite vegetables, they are more likely to give broccoli a high rating (see p. 109). Also, it's a good idea to eat vegetables during pregnancy, as taste preferences begin to develop in utero.

Combining a new flavor with a well-liked familiar flavor is another powerful way to build a new preference. Researchers have found that the two tastes cannot be given more than nine seconds apart. You can try this approach with your baby: simply mix the two flavors together. For example, babies develop a preference for pure carrot juice after having it mixed with milk. Sweeteners work too. A common way to introduce solid food to babies is to puree it in a blender, an approach that lends itself very well to mixing-and-matching of flavors. This pattern persists throughout life. College students who are given broccoli with sugar will later rate broccoli alone as being more pleasant than cauliflower alone—and will do the converse if they are given sweetened cauliflower. Coffee drinkers often initially add sugar or milk but eventually take it black. Learning can even be negative: pairing a flavor with bitter quinine, found in tonic water, can reduce liking.

One approach we don't recommend is offering dessert as a reward for finishing dinner. The urge to consume foods that contain calories is a powerful motivator, as confirmed both by our everyday experience and by behavioral experiments. But when kids eat dessert right after a meal, something odd happens: their preference for foods eaten earlier decreases. Why?

Recall that our brains want us to like foods that are high in calories. And

the gut detects calorie content many minutes after we eat. So, because the calories from foods eaten earlier are still being processed when dessert arrives, the earlier foods actually encourage a preference for the taste of dessert rather than the reverse.

One solution to this problem would be to give dessert before the new flavor—ideally within nine seconds. There's a converse problem if you give dessert too early: your kid's not likely to be hungry come spinach time. If you're going to use this approach, we recommend serving dessert more than thirty minutes after the meal—or offering a bite or two right at the time of the meal!

The developing brain's first teachers for food preference are the tongue (taste) and gut (nutrient content). In the tongue, molecules that trigger one of the five basic taste sensations bind to receptor molecules in a taste bud, which then generates a chemical signal inside the sensory neuron that bears the receptor molecules. Each neuron, taste bud, and axon is a little communication line that is assigned the identity of *sweet*, *salty*, and so on by virtue of where it connects. These labeled lines convey basic information about chemicals found in foods.

Scientists used to think that sweetness taste buds were found only in one part of the tongue, but now we know that each type of taste bud is found all over the tongue. Receptor cells form in the tongue very early, during week eight of gestation. By week thirteen, taste buds are present throughout the mouth and are connected to nerves that go to the brain. By the time the axons are functionally connected to brain structures, they are hooked up and interpreted correctly.

Taste signals are translated into electrical impulses that are conveyed along the neurons' axons to the *nucleus solitarius*, a cluster of brainstem neurons (see figure, p. 106). The nucleus solitarius is an important station not just for taste information but also for other visceral signals, including the presence of fat in food. Other organs besides the gut also send signals: the cardiovascular system, liver, and lungs. The nucleus solitarius's many jobs include generating the gag reflex, coughing, and responses relating to breathing, digestion, and the heart. A notable reflex in this general category is the *gastrocolic response*, in which the eating of food, especially if it is fatty, helps trigger defecation after about half an hour

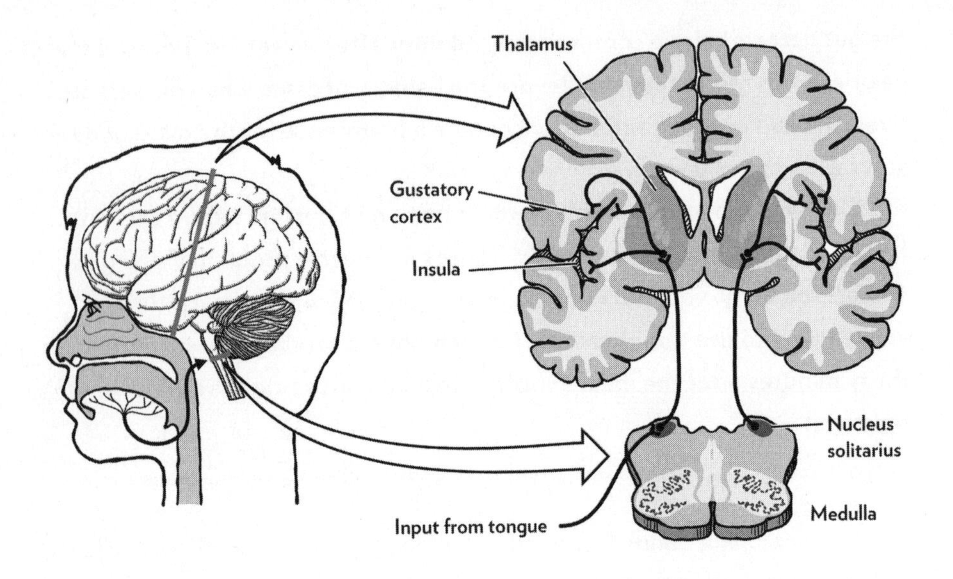

(in babies and grown-ups). This reflex can be useful in making diaper-change time more predictable.

These labeled lines for taste and calorie content serve an important evolutionary purpose. Starting early in life, everyone strongly prefers sweet tastes and calorie-rich foods. The benefit of a sweet tooth is obvious: sweet things are usually loaded with valuable calories. Glutamate is similarly useful, as is salt. Bitter and sour tastes are rejected. In all cases, these experiences stimulate powerful teaching signals that tell the brain that something good or bad has happened. These teaching signals are conveyed through the nucleus solitarius onward to the thalamus, striatum, and neocortex.

DNA does not just encode receptors; it also carries instructions telling each cell which receptor protein it should make. One example that has been studied carefully is taste neurons that detect sweetness. These neurons all make the same sweet receptor molecule, a protein dedicated to the job of binding to sugar. Researchers have applied genetic engineering techniques to make mice that contain the DNA sequence for a receptor for morphine (which mice normally can't taste) in the exact location where the sequence for the sweet receptor would normally be found. Mice with this genetic alteration can't taste sugar, but they can taste chemicals related to morphine—and consume them avidly, as if they were sweet. This is true even for amounts of a morphinelike drug that are far too small to be addictive. This result suggests that the surrounding DNA acts as a signpost that the receptor at that location should be hooked up to brain wiring that carries the message "It's sweet, and I want more."

In contrast to having just one sweet receptor, our genome has multiple receptors for other tastes—including dozens that are sensitive to bitter chemicals. Babies instinctively regard bitter flavors as unpleasant. Many toxic chemicals are bitter and are often found in plants. Bitterness is a natural signal to get us to spit these things out, but paradoxically, we can teach our brains to enjoy bitter flavors. This flexibility is useful, since it allows your child's brain to adapt to a food that has nutritional value but happens to contain one of the chemicals found in other bitter, less healthy substances that occur naturally. With practice, we can learn to like tonic water, coffee, and broccoli—and children should like broccoli (see *Practical tip: Getting your child to eat spinach*, p. 104). We train our brains to accept particular bitter flavors by using rewards (pairing them with liked flavor, calorie content, or social approval) and reject other bitter flavors using penalties (if they lead to physical illness or social disapproval).

Learned preferences for food accumulate over the first few years of life. A liking for salt emerges by age two, and eventually for more complex flavors such as cherries. This process starts before birth. Indeed, preferences for the foods of one's own culture may be transmitted through mother's milk and even in the womb.

The womb is a flavorful environment. Fetuses swallow cupfuls of amniotic fluid per day, and the fluid can reach their olfactory epithelia. As we have said, at birth, newborns have a pronounced preference for the flavor of amniotic fluid (especially their own) over water. Amniotic fluid can also carry traces from the mother's diet. To test for flavor conditioning in utero, researchers had pregnant women drink about ten ounces (thirty centiliters) of carrot juice every other day for three weeks during the third trimester. Their babies were less likely to make faces when given carrot-flavored cereal for the first time—and more likely to eat it.

Flavor preferences learned in infancy can last for years.

Flavor learning in utero has been shown in other species of mammals, not just humans. Rabbit pups show an increased preference for juniper berry flavor after prior exposure to this taste in their mother's milk or in the mother's diet during gestation, or even if they are housed with juniper-scented fecal pellets.

PRACTICAL TIP: WORRIED ABOUT YOUR CHILD'S WEIGHT?

As a parent, it's normal to be concerned about what your child eats, but anxiety about childhood obesity is usually neither necessary nor helpful for most children. Parents need to be careful that the solution they choose doesn't end up causing more harm than the problem.

Because they're growing, children need more calories than adults per pound of body weight. Young children also need to eat a lot more fat than adults do. The National Institutes of Health recommends that children under two should get 50 percent of their calories from fat, and children older than two should get 25–35 percent of calories from fat. Dietary fat is an essential contributor to early brain development. Food restriction is tricky in children: low-calorie diets may prevent them from growing properly, and low-fat diets are associated with inadequate intake of important nutrients. Complicating things further, it's difficult to be certain how much a child should weigh at a given age because kids often put on some extra fat in preparation for a growth spurt.

Given these difficulties, how do you decide how much food your child should be eating? In most cases, if you provide a variety of healthy foods and opportunities for regular exercise, you can trust your child's brain to manage food intake correctly. (This approach probably will not work as well if you have a lot of junk food in the house.) The brain's weight-regulation system is extremely complicated, including more than twenty molecules that increase eating and a similar number that decrease eating, based on a variety of nutritional cues. The brain will do its best job of balancing food intake with energy use if children are simply allowed to eat when they are hungry and stop when they are full. This approach also helps them learn to regulate their own eating as adults without binging or starving.

Growing up in a weight-obsessed culture is particularly difficult for girls. Healthy puberty requires the addition of body fat, in the form of breasts and hips, at just the age when girls are particularly sensitive to body image problems (see chapter 8). Eating disorders affect more than one in a hundred young women between fifteen and twenty-four years of age. (There is roughly one man with an eating disorder for every eight women.) In one

longitudinal study, teenage girls who reported dieting repeatedly or being teased about their weight by family members were much more likely than average to have developed an eating disorder or to have become overweight (from a normal starting weight) five years later, suggesting that strict parental efforts to control their daughters' weight were typically ineffective—or even counterproductive. Girls who reported eating regular family meals in a pleasant atmosphere were less likely than average to develop an eating disorder or become overweight.

Pups seem to have multiple ways of learning about what might be okay to eat. It's at least possible that some distinctive aspects of sushi flavors, such as fish and seaweed, could have reached Sam's daughter before birth or in infancy.

Taste preferences can also arise indirectly. For instance, four-month-old infants are less likely to make faces when presented with salty water if their mothers experienced early pregnancy nausea. We don't know for certain why morning sickness would lead to a liking for salt. One possibility is that sick mothers become dehydrated or depleted of sodium, leading to the secretion of the signaling molecules *renin* and *angiotensin*, which cause a desire for salt in adults. (Renin is an enzyme, secreted by the kidney, that participates in regulating blood pressure; angiotensin is a **peptide** that causes blood vessels to constrict.)

Considerable information about flavors is also transferred through mother's milk. One aspect of children's experience is simple familiarity, leading to tolerance. If a lactating mother consumes garlic oil or carrot juice, babies are less likely to react, either positively or negatively, when the same flavoring is added to a bottle.

Flavor preferences learned in infancy can last for years. To illustrate, let's look at an experiment that focused on baby formula, which comes in several varieties based on milk, soy, and protein hydrolysate (for babies with allergies to milk and soy). Nonmilk formulas are notably sour and bitter, and the hydrolysate formulas are particularly nasty. (Try one sometime.) Generally, just consuming a food multiple times is sufficient to reduce negative reactions. Furthermore, infant taste is particularly malleable during the first few months. In this experiment, children who were fed soy or hydrolysate formula before the age of one still chose to drink it at age four or five over milk-based formula (as opposed to milk-reared kids, who refused soy- and hydrolysate-based formulas and made yucky faces). Soy- and hydrolysate-fed kids were considerably more tolerant of sour or bitter flavors added to apple juice. They also were more positive about other bitter foods—including broccoli.

As children become more verbal, they acquire additional ways of learning about flavors and smells. In both children and adults, perception and liking can be influenced by a word label. For example, adult subjects gave a more pleasant smell rating when they rated a container labeled *cheddar cheese* compared with a container labeled *body odor* even though the containers emitted the same odor. When the adults' brains were scanned, activity was changed in the **rostral anterior cingulate** cortex, the **medial** orbitofrontal cortex, and the amygdala. These regions receive information from both smell and taste pathways and appear to be places for making mental associations with flavors. As children grow, they learn to make associations between foods and complex contexts, including social value, as adults do. Pleasure can be a force for good in your child's development, an idea that we explore in the next section.

PART FOUR

THE SERIOUS BUSINESS OF PLAY

THE BEST GIFT YOU CAN GIVE: SELF-CONTROL

PLAYING FOR KEEPS

MOVING THE BODY AND BRAIN ALONG

ELECTRONIC ENTERTAINMENT AND THE MULTITASKING MYTH

Chapter 13

THE BEST GIFT YOU CAN GIVE: SELF-CONTROL

AGES: TWO YEARS TO SEVEN YEARS

At the age of three or four, resisting temptation is always a visible struggle. Finding a good strategy is a key element to success—and the good news for parents is that strategies can be taught. Indeed, learning self-control strategies at an early age can pay off for years afterward.

Young children's play contributes to the development of their most important basic brain function: the ability to control their own behavior in order to reach a goal. This capacity underlies success in many areas that parents care about, from socialization to schoolwork. The neural circuits that are responsible for it are some of the latest-developing parts of the brain, as any parent of a two-year-old can tell you. But even at this age, you can detect and encourage an early stage of their mental growth, the inhibition of behavioral impulses.

Preschool children's ability to resist temptation is a much better predictor of eventual academic success than their IQ scores. The classic test for this ability, devised by psychologists, is to put a marshmallow on the table and tell the child that she can have two marshmallows if she can wait a few minutes without eating the first one. Alternatively, she can ring a bell at any time to bring the researcher back into the room and get just one marshmallow. The average delay time is about six minutes for a four-year-old. A child who can hold out fifteen minutes at that age is doing exceptionally well and definitely deserves both marshmallows without further delay.

More than a decade later, those preschool delay times correlate strongly with adolescent SAT scores—predicting about a quarter of the variation among individuals in one study. Delay times in preschool also correlate with the ability to cope with stress and frustration in adolescence, as well as the ability to concentrate. Other tests of the ability to inhibit behavior for delayed gratification correlate with math and reading skills early in elementary school, which makes sense considering that learning academic subjects requires concentration and persistence.

As it improves with age, self-control continues to predict academic success. In a study of eighth graders, self-discipline at the beginning of the school year—measured in part by the students' ability to carry a dollar for a week without spending it in order to earn another dollar—predicts grades, school attendance, and standardized achievement test scores at the end of the year. The students'

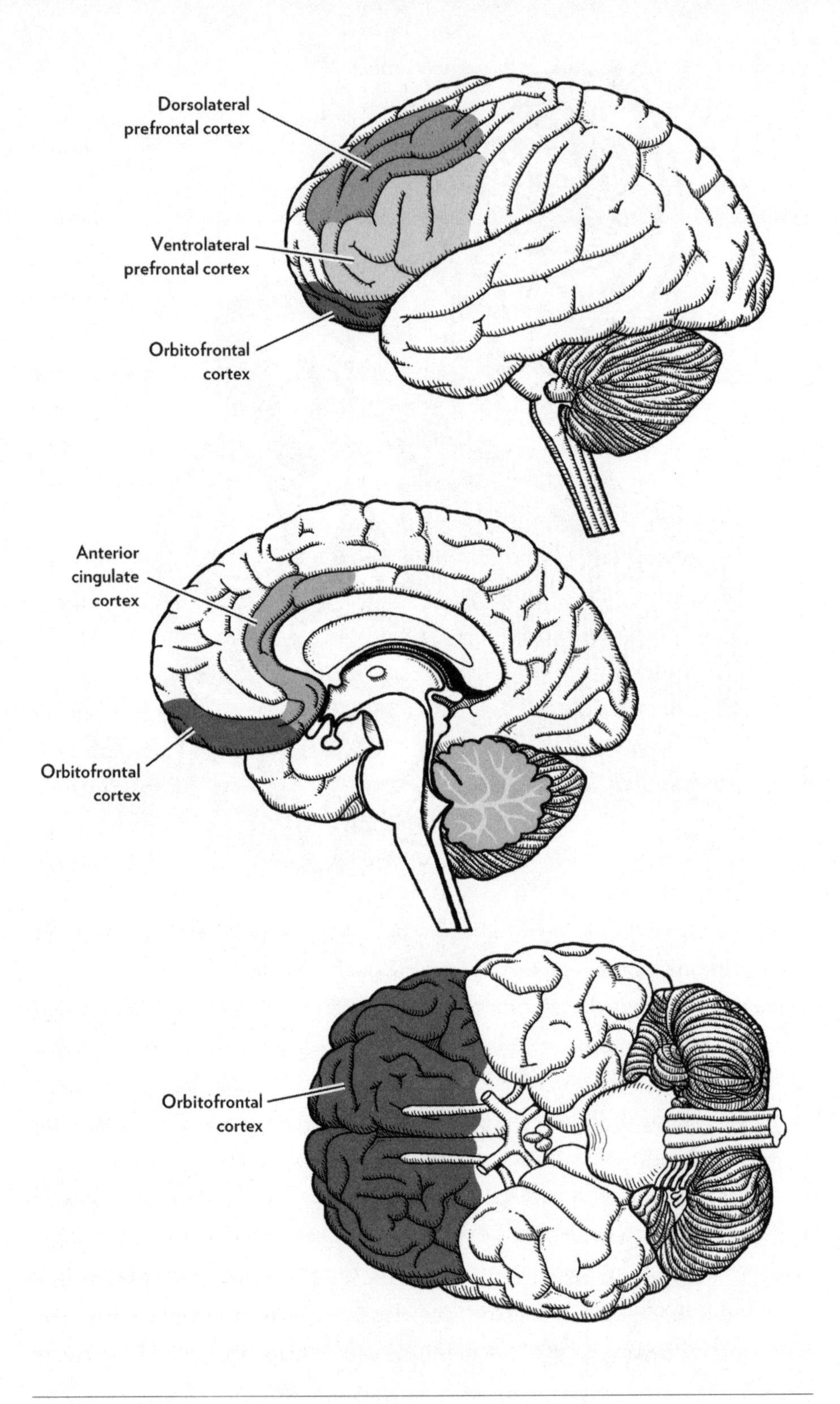
Dorsolateral
prefrontal cortex
Ventrolateral
prefrontal cortex
Orbitofrontal
cortex
Anterior
cingulate
cortex
Orbitofrontal
cortex
Orbitofrontal
cortex

self-control ability accounted for twice as much of the variation among individuals on all these measures as their IQ scores.

The ability to regulate your own behavior is important not only for academic achievement but also for interpersonal success. Children who are skilled at behavioral self-control show less anger, fear, and discomfort and higher empathy than their peers of the same age. Other people rate these children as more socially competent and more popular, even years later, probably because they are good at regulating their own emotions and taking the emotions of others into account.

Some amount of brain maturation is a necessary first step in the development of self-control. Around ten months, in the earliest sign of this capacity, babies are able to choose which aspects of the environment to select for attention (as opposed to paying involuntary attention to whatever suggests itself; see chapters 10 and 28). Once this occurs, they can comfort or distract themselves and sometimes change their behavior instead of persisting on a course of action that's not effective, as younger infants do. The more complex ability to deliberately inhibit behavior, control impulses, and plan actions, called *effortful control*, is first seen around twenty-seven to thirty months of age. It allows children to do things like remember to use their inside voice when they are excited or keep their hands out of the cookie jar (at least some of the time). Toddlers develop the ability to inhibit behavior on command between their second and third birthdays. Effortful control then improves rapidly until the fourth birthday and more slowly through age seven. Resistance to distraction continues to get better throughout childhood, reaching adult values in the middle to late teens.

Two related skill sets depend on similar brain regions and thus tend to develop in parallel with effortful control. One is *cognitive flexibility*, the ability to find alternative ways to achieve a goal if the first attempt does not succeed, and to adjust behavior to fit the situation, like not running near the pool. The other is *working memory*, the ability to remember task-relevant information for a short period of time, such as recalling which solutions to a puzzle you already tried. Together, these three abilities are collectively called *executive function*.

As these abilities grow with age, children become progressively better at sequencing behavior appropriately, keeping track of multistep tasks, and resisting or recovering from distractions. Executive function, which provides the core ingredients for self-control in adulthood, depends on the **prefrontal cortex** and the anterior cingulate cortex. The prefrontal cortex shapes behavior in pursuit of goals by activating or inhibiting other brain regions. The anterior cingulate is activated by tasks that require cognitive control, particularly monitoring and detecting errors in performance and deciding among conflicting cues. Another part of the anterior cingulate is connected with the orbitofrontal cortex, hippocampus, and amygdala and is involved in regulating emotions.

One measure of a child's growing executive function is the strength of his cognitive control. This ability is typically measured by the *conflict cue task*, in which children are asked to rapidly detect the appearance of a target. This task is easier (and so performance is faster) if the child knows in advance where the target will appear. In a simple version of the task (see figure below), the target is preceded by one of two cues, a row of five fish that all face toward the target's future location (*congruent condition*), or another row with the center fish facing the target location and the other four fish facing the wrong way (*incongruent condition*). Response times are slower, for children and adults, when faced with incongruent conditions, because the brain must inhibit the automatic tendency to follow the majority of cues and instead concentrate on the center fish alone. Performance on this test correlates with parents' evaluations of their children's self-control ability.

Congruent

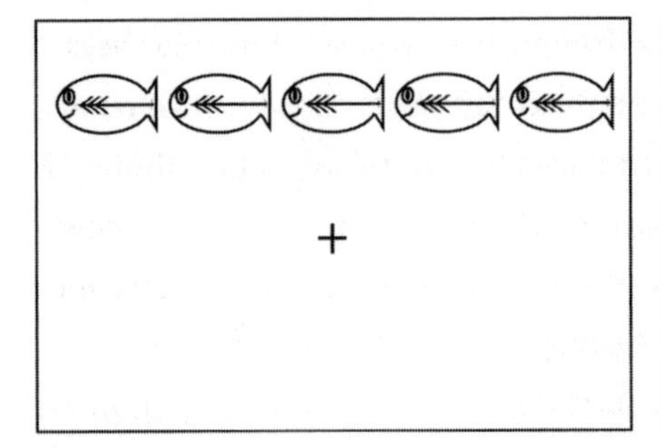

Incongruent

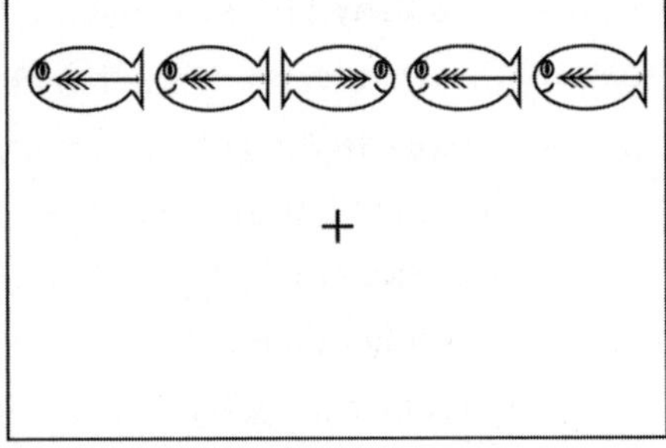

Stable individual differences in these measures become evident during early childhood. The ability to self-regulate is moderately heritable; some evidence suggests that this has to do with the genes that regulate the neurotransmitters dopamine and **serotonin**. But children's experiences also have an important effect on their self-control in studies that take genetic influences into account.

Four-year-olds who do well on the marshmallow task typically distract themselves from thinking about the tempting object during the delay period. They cover their eyes, turn their backs on the marshmallow, or try to think about

PRACTICAL TIP: IMAGINARY FRIENDS, REAL SKILLS

An average four-year-old child who is asked to stand still for as long as possible can manage it for slightly less than one minute. If he's asked to pretend he is a guard outside a castle, though, he can hold his pose four times that long. The reason is simple: pretending is fun, which means it's easy to get children to participate enthusiastically. Children can choose games that reflect their own interests, so they can supply their own motivation for achievement, which teaches self-control—an ability that is more useful in adulthood than modifying your behavior to please others. Most important, imaginative play has rules that children take seriously. To play school, you have to act like a teacher or a student, and inhibit your impulses to act like a fighter pilot or a baby. Following these rules provides children with some of their earliest experiences with controlling their behavior to achieve a desired goal.

An innovative preschool program called Tools of the Mind has achieved remarkable success using the power of play to teach disadvantaged children how to control their impulses and organize their behavior in pursuit of goals. Children are asked to plan how they want to spend their play time ("I am going to take the baby to the doctor") and then to stick to that plan. Teachers use physical reminders to help children regulate turn-taking, such as having one child hold a cardboard ear and another child a mouth to remind them who is supposed to be listening and who is supposed to be talking right now. The goal is to help children develop the ability to control their own behavior, rather than simply following rules to gain gold stars or avoid time-outs.

The program is only a few years old, but the early data are exciting. In one study, done in a low-income neighborhood, more than twice as many Tools of the Mind students could successfully perform a difficult attention-demanding task, in comparison to those who were in another preschool program. We look forward to future research aimed at finding out how long these self-control gains last, whether they improve later academic performance, and which of the techniques used in the program are most effective. In the meantime, we encourage parents to adopt some of these techniques to help their three- and four-year-olds learn to regulate their own behavior through imaginative play.

PRACTICAL TIP: **LEARNING TWO LANGUAGES IMPROVES COGNITIVE CONTROL**

Becoming bilingual gives children cognitive advantages beyond the realm of speech. Learning multiple languages is challenging in part because the person must direct attention to one language while suppressing interference from the other. This interference causes bilingual people to be slower at retrieving words and have more "tip-of-the-tongue" experiences than monolingual people.

There are benefits to meeting these challenges. Bilingual children outperform monolingual children on tests of executive function. Before their first birthday, bilingual children learn abstract rules and reverse previously learned rules more easily. They are less likely to be fooled by conflicting cues, such as a color word like *red* written in green ink, which psychologists call the Stroop task. This pattern continues into adulthood and even shows up in nonverbal tasks. Selecting appropriate behavior in two different languages seems to strengthen bilingual children's ability to show cognitive flexibility according to context—an aspect of self-control.

Bilingual children also outperform monolingual children on tests that measure the ability to understand what other people are thinking (see chapter 19). This advantage may develop because bilinguals get more practice at taking the perspective of other people, since they need to choose the appropriate language for the person they're talking to.

Bilingual people may exert cognitive control not only better, but by using different brain areas. During a task that requires them to resolve differences between two conflicting sources of information, bilingual children's brains show activation not just in the portion of the prefrontal cortex that everyone uses for conflict resolution, but also in **Broca's area**, the speech region that processes grammatical rules.

Bilingualism may also protect the brain from cognitive decline in aging. People who have spoken two languages actively for their entire lives experience the onset of dementia four years later, on average, than their peers who spoke only one language.

That's a lot of advantages—and we haven't even mentioned the most important use of a second language, to communicate with people.

something else. One of the tricky aspects of self-control is that it has a certain circular quality: it takes discipline to learn discipline. Children who can focus on a task without giving in to distractions are also going to be better at improving this ability through practice. The key is for parents and teachers to provide scaffolding to support the learning process until the child's self-control is strong enough to stand on its own (see chapter 29). This process is easier, especially in young children, if a child finds the task rewarding. So keep an eye out for age-appropriate, multistep projects, like making art or building something, that your child enjoys.

Here's some even better news. Improvement of self-control is not limited to a sensitive period (see chapter 5). We all know that even in adults, the ability to control your own behavior is limited, but it can be increased by training. As the psychologist Roy Baumeister puts it, willpower is like a muscle: the more you use it, the better it works. His studies show that self-control can be improved by practicing any sort of self-control, from dieting to money management to brushing your teeth with your left hand, on a regular basis. College students who did these exercises for several weeks reported improvements in their ability to complete a variety of tasks requiring self-control, from going to the gym regularly to managing money to doing housework. Indeed, the legendary tough parenting style of many Asian parents appears to be directed specifically at instilling self-discipline, which could account for the high achievements of their children.

Succeeding at challenging self-control tasks builds more success, but repeated failure may instead teach the child that there's no point in trying.

This plasticity in adulthood strongly suggests that children should also be able to increase their self-control abilities through experience. In young children, warm supportive mothering is associated with improved self-control ability, even when the mother's genetic contribution is factored out. During elementary school, computer-based attention training can increase reasoning abilities, and children with the poorest attention before training show the most benefit.

Structured play with other children can also improve executive function. One promising preschool program uses play to improve self-control in children from disadvantaged backgrounds (see *Practical tip: Imaginary friends, real skills*, p. 117). All these forms of training probably improve performance of the brain network involved in cognitive control.

You can help by encouraging your child to exercise as much self-control as possible in the context of enjoyable experiences like playing board games. The rules of the game require your child to resist such impulses as moving his piece when it's not his turn. If you stand over your child and manage every step of the process, you're depriving him of the experience that will allow him to learn how to organize his own actions. On the other hand, if your child consistently fails to control himself, the game may be too difficult for his developmental stage. Succeeding at challenging self-control tasks builds more success, but repeated failure may instead teach the child that there's no point in trying.

Children who fail to develop an age-appropriate level of self-control can end up with troubles that multiply. In school, they are likely to get poor grades because they have difficulty concentrating and completing assignments. In addition, parents, teachers, and other students are likely to find them difficult to handle. Because punishment is more likely to induce resentment than to increase self-control (see chapter 29), its repetition can eventually establish the poorly self-regulated child as the "bad" or "rebellious" member of the class or of the family. This image, in the eyes of teachers, parents, peers, and the child, can outlast the original problem and contribute to a downward spiral into poor performance and delinquency.

Even young children can benefit from spending time directing their own activities, particularly if they are engaging in imaginative play, alone or with other children. Parents can help by encouraging children to pursue their own interests and enthusiasms. Some children are captivated by producing music, while others would rather build a castle. It probably doesn't matter exactly what excites your children; as long as they are intensely engaged by an activity and concentrate on it, they will be improving their ability to self-regulate and thus their prospects for the future. Play has many benefits for children, as we will explore in the next chapter.

Chapter 14

PLAYING FOR KEEPS

AGES: TWO YEARS TO EIGHTEEN YEARS

For Pigface, life at the zoo had recently improved. Previously prone to clawing and biting himself, he had been given sticks and other items, with which he developed various tricks. One was to push a basketball with his snout and sometimes snap at it. Hoops of hose were favorite objects, which he would nose and chew and sometimes swim through. On days when his tank was being cleaned, he would get in front of the stream of incoming water and remain there unmoving, just feeling the water run over his face. Once the refilling was done, he was off again. These activities make Pigface sound like an otter or a seal. Movies of him certainly look like such frolicking—except in very slow motion. The behaviors only look like mammalian play when played back at three times normal speed, because Pigface is a turtle (see photos, p. 122).

Similarities run deep between how we and other animals play. When you watch your children play, you may think it's cute—but they are also learning to match their capacities to future life needs. Object play such as pushing a truck or ball around resembles later activities such as wielding a hammer, hunting, or building machinery. Indeed, play can also take a distinctly animal-like quality. Babies and toddlers like to chew and pull things apart. Biting is a common nuisance in preschool centers—and it's also practice for the kinds of predation that our species once depended on for food.

Play is widespread among animals, beyond the familiar cases, mammals and birds, to vertebrates and even invertebrates. How can we be sure that an animal is playing? Researchers use three criteria. First, play resembles a serious behavior, such as hunting or escaping, but is done by a young animal or is exaggerated,

awkward, or otherwise altered. Second, play has no immediate survival purpose. It appears to be done for its own sake and is voluntary or pleasurable. Third, play occurs when an animal is not under stress and does not have something more pressing to do.

These criteria for play are met by leaping needlefish, water-frolicking alligators, and lizards. At the National Zoo in Washington, D.C., monitor lizards play games of keep-away. The largest monitor species, the Komodo "dragon" lizard, plays tug-of-war with its keepers over a plastic ring. It can pick a familiar keeper's pockets of notebooks and other objects, then walk around carrying them in its mouth. A movie of a playing Komodo dragon looks quite a bit like play with a dog, only slowed down to about half speed.

The behavior is not just displaced foraging or hunting. If the ring is coated in tasty linseed oil or animal blood, playfulness vanishes and is replaced by a pronounced possessiveness. YouTube videos of Komodo dragons swallowing whole pigs—or even other Komodo dragons—suggest that these food-oriented behaviors are not easily confused with play.

The fact that play is so widespread suggests that it arose long ago in the history of animals. It appears in many animals with far less social complexity than people. This universality suggests that even though play is literally fun and games, it must serve some vital function. In other words, when your child is playing, he is doing something crucial for his development. Furthermore, the features of his play are distinctive not only to him but to humans in general.

Play takes different forms in different animals, including humans. Its content provides some hint as to what it might be good for. Play researchers (there's a fun-sounding job) recognize three major types of play. Most common is *object play*; that's what Pigface does with basketballs and hoops. Object play is typically found in species that hunt, scavenge, or eat widely. About as common is *locomotor play*, such as leaping about for no apparent reason. (The term *locomotor* has to do with coordinated movement through space, such as crawling, walking, or running.) Locomotor play is common among animals that move around a lot, for instance, those that swim, fly, or live in trees—and notably, often must get away from predators. The third and most sophisticated form of play is *social play*. Social play can take many forms, including mock fighting, chasing one another, and wrestling. Pretending is a major component of social play.

Social play is especially prominent in animals that show a lot of behavioral flexibility or plasticity. (And of course the most flexible, plastic young animal of all is your child.) In mammals and birds, this boils down to a simple rule: if your species has a big brain for its body size, you probably engage in social play. Among these species, most of the variation in brain size occurs in the

DID YOU KNOW? PLAY IN ADULT LIFE

In many people, play continues in adulthood and is a major contributor to problem solving. Physical scientists often report having built and taken apart machines when they were kids. Sam did this too, with activities such as building a sundial in elementary school.

Conversely, work in adult life is often most effective when it resembles play. Indeed, total immersion in an activity is often the mark that an activity is intensely enjoyable—the concept of *flow*, or what athletes call *being in the zone*. Flow occurs during active experiences that require concentration but are also highly practiced, where the goals and boundaries are clear but leave room for creativity. It is most likely when your abilities are just sufficient to meet the challenge, in tasks that are neither easy nor impossible. This describes many adult hobbies, from skiing to music, as well as careers like surgery or computer programming. Such immersion can make solving a great challenge as easy as child's play. Encouraging your child to pursue tasks that produce flow is a great way to contribute to his lifelong happiness.

forebrain; different mammals or birds of a given body size will have about the same amount of brainstem, but very different amounts of neocortex (in mammals) or forebrain (in birds). Animals with more neocortex or forebrain typically live in larger groups and have more complex social relations. Ducks engage in "coordinated loafing," in which they basically just hang around together. Great apes and their relatives (such as humans) form societies in which alliances are constantly shifting, and in which the young play recognizable games, such as chasing, wrestling, and tickling. In doing these things, your child is indulging her inner ape.

This phenomenon is not limited to vertebrates. Among invertebrates, perhaps the most complex brains are found in cephalopods, which include squid and octopus. Octopuses use their water jets to push floating objects like pill bottles back and forth in a tank or in a circular path. Despite this behavioral complexity, octopus brains are still small by vertebrate standards—half the diameter of a dime, smaller than those of even the smallest mammals. Another invertebrate that appears to play is the honeybee, which has one of the largest and most complex nervous systems among invertebrates. As a counterexample, playlike behavior is not reported in houseflies.

Now, maybe play isn't "for" anything. Perhaps play behavior is simply early maturation, precocious behavior that develops before it is absolutely required. Another possibility is that play is what our brains do when there are no more pressing matters—a screensaver for the mind, as it were.

But these ideas are contradicted by one key piece of evidence: play is fun. At first, this may seem like an odd argument. Aren't fun activities the ones we engage in for their own sake? Superficially, yes, but dig a little deeper. The enjoyment of an activity is a survival trait. We are wired to like activities that are helpful for our survival. For example, we may think we seek sex because it's fun, but in reality, sex is essential. Sex is fun because seeking it is adaptive. People who don't like sex have a harder time finding mates and having kids. In general, enjoying an activity is a hardwired response that causes our brain to seek out that activity. If these essential behaviors weren't enjoyable, we might forget to do them, and then we wouldn't make it through life very well. On these grounds, it seems that play must have an adaptive purpose, providing some survival advantage.

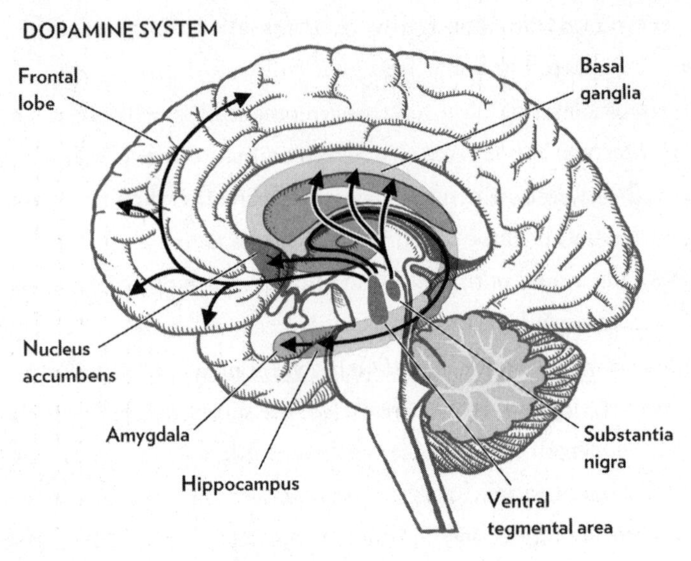

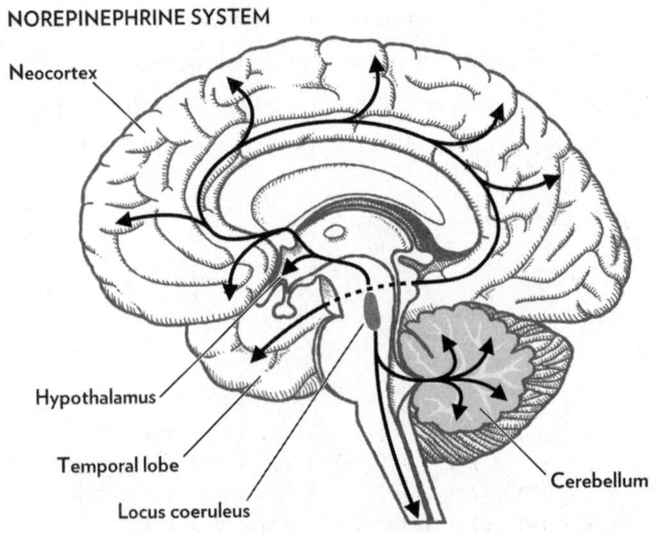

The brain generates chemical signals that encode a key component of fun: *reward*, the quality that makes us come back for more. Reward is conveyed within the brain by dopamine, a neurotransmitter that has many functions depending on where and when it is secreted. Dopamine is made by cells in the brain's core, in the **substantia nigra** and **ventral tegmental area** (see figure above). In rats, dopamine and play are linked. Among chemicals that activate receptors for various neurotransmitters, only a few increase play behavior, including drugs that activate dopamine receptors.

One way to find out what play is good for is to take it away from animals and see how they fare. The problem is that this experiment is nearly impossible to do. Animals (including children) are irrepressible; they play under the most adverse of conditions. The only way to get an animal to stop playing is to restrain its mobility. This severe restriction leads to decreases in physical activity and increases in stress, as measured by the amount of the stress hormone cortisol in saliva. Play, exercise, and stress are closely linked.

Though the deprivation experiment is hard to do, that very fact means that seeing an animal play already tells us something good about its state. In young squirrel monkeys, lower levels of cortisol are associated with high amounts of play, suggesting either that play reduces stress, or possibly that unstressed monkeys are more likely to play. In bear cubs during their first year of life, survival over the winter is highly correlated with the amount that cubs played during the preceding summer. This suggests that play might be an indicator of health or resistance to stress. No matter how you slice it, seeing your child play is a good sign.

Play is necessary for forming normal social connections.

Play activates other brain signaling systems as well, including the neurotransmitter norepinephrine (see figure opposite). Its close relative epinephrine (also known as adrenaline) is released to the body as a fast component of stress-related signaling (see chapter 26). As the main activator of the sympathetic nervous system, epinephrine mobilizes our energies for "fight, flight, or fright," as the medical-school mnemonic goes. Norepinephrine too is involved in rousing us to attention and action, but by acting as a neurotransmitter.

Norepinephrine also facilitates learning mechanisms at synapses. In some neurons, norepinephrine improves brain plasticity, so that change becomes possible when this chemical is present in elevated amounts. The same is true for dopamine, which accounts for how reward leads to long-term changes to make us want more—brains are rewired when reward occurs.

Though real-life stressors trigger the release of both epinephrine and cortisol, play does not increase cortisol. This hormone helps us in genuinely dangerous situations by shutting down functions that are dispensable for a little while, such as learning of experiences that are not immediately related to the stressful context (see chapter 26). It is safe to say that if you find play to be a source of stress, you're not doing it right. Even violent video games, which raise physiological arousal as measured by epinephrine-based response, do not increase cortisol. In some cases, cortisol levels actually decrease—people work off stress by shooting 'em up. On the whole, play is associated with responses that facilitate learning.

The conditions of play—the generation of signals that enhance learning without an accompanying stress response—allow the brain to explore possibilities and learn from them. In other words, a major function of play may well be to provide practice for real life. As we've written before, the use of a skill or other mental capacity builds up that ability. Evidence from animals suggests that this is the case for play, which usually reflects an animal's more serious needs. Kittens play at pouncing on things, a behavior that resembles the hunting they do later. Fawns don't pounce much, but they do gambol around, a behavior that resembles escape.

Risk taking in children's play may be an important developmental process. It tests boundaries and establishes what is safe and what is dangerous.

So it's possible that play is practice that prepares animals for the real activity later—when it matters. For example, in chapter 13 we described Tools of the Mind, a preschool program that uses complex play to get children to make elaborate plans and exercise self-restraint—practice for the prefrontal cortex and self-control. Even before that the kindergarten movement, which in the nineteenth century popularized the concept of preschool education, was based on the idea that songs, games, and other activities were a means for children to gain perceptual, cognitive, social, and emotional knowledge that would prepare them for entering the world.

In mammals, play is necessary for forming normal social connections. Rats and cats that are raised in social isolation become incompetent in dealing with others of their kind, typically reacting with aggression. In our species, dysfunction in adults is often presaged by abnormal play as children. A notable feature of psychopaths is that their childhoods were lacking in play. Serial killers are often reported to have had abnormal play habits, keeping to themselves or engaging in notably cruel forms of play. Sometimes such problems are associated with early-life head injury. (Needless to say, abnormal play does not necessarily mean that you are raising a serial killer.)

Culture is also transmitted through play. Middle-class mothers in the U.S. encourage their infants to pay attention to objects and are likely to prompt them to play with toys such as blocks. In contrast, Japanese mothers encourage their babies to engage in social interactions while playing, for example, by suggesting that they feed or bow to their dolls. Communities that emphasize the development of independence place more importance on object play, while interdependent communities encourage social play.

There are some downsides to play, too. For one thing, though play is defined as occurring in the absence of stressors or external threats, children aren't always good at detecting threats, so play can be dangerous. People are not alone in having this problem. In a study of baby seal mortality, twenty-two of twenty-six deaths resulted from the pups playing outside the sight or sound of their parents. Play can distract people and other animals from recognizing danger.

But even here, play may be practice for real life. Risk taking in children's play may be an important developmental process. It tests boundaries and establishes what is safe and what is dangerous. In the U.S., playground equipment has been made very safe, leading to the unanticipated problem that children lack experience with such boundaries, which may lead to trouble later in life.

In addition to providing experience, play also helps children learn what they like and don't like. The Nobel chemistry laureate Roger Tsien tells of reading about exotic chemical reactions before he was eight years old, then trying out the reactions for himself. He was able to get beautiful color changes to happen in his house or backyard. Because he didn't have enough laboratory glassware, he had to make equipment from used milk jugs and empty Hawaiian Punch cans (see photo, p. 130).

After he grew up, Roger won the Nobel Prize for developing colored dyes and proteins that become brighter or change hue when they encounter chemical signals in living cells, including neurons. (This can make certain biological processes, such as brain signaling, much easier to see and understand.) The invention of tools to visualize what is happening in active biological systems, Roger's great contribution to science, had its roots in that childhood interest in home chemistry experiments. Not all childhood experiments lead to such heights, of course, but no matter what your children's eventual interests may be, discovering them may be one of the most important outcomes of play.

Think of it this way: play is the work of children. It is perhaps the most effective way for them to learn life skills and to find out what they like. For these reasons, it is important to prevent play from becoming a compulsory, dreary activity, as its enjoyable nature is part of what makes most children grow like dandelions. So, rather than trying to change your child's personality through enforced activities (which is tough anyway; see chapter 17), let her play, and help her become who she's going to be.

A homemade chemistry apparatus built by later Nobel laureate Roger Tsien when he was about fourteen years old. He was trying to synthesize aspirin.

Chapter 15

MOVING THE BODY AND BRAIN ALONG

AGES: FOUR YEARS TO EIGHTEEN YEARS

It's obvious that your child is using her brain when she's reading, but you may not realize that her brain is also getting a workout when she plays soccer. Sports and exercise in childhood are beneficial to motor control and cognitive ability—and, perhaps most important, they set up habits that can keep your child's brain healthy into old age. For all these reasons, the recent trend to reduce or eliminate physical education in U.S. schools is likely to be bad news for children.

The control of movement and balance continues to improve throughout childhood. During this time, the brain gets better at coordinating its commands to various muscles and interpreting feedback from sensory systems. Skills like walking over rough ground may feel effortless to you, but your brain is working very hard to produce that impression. Controlling the body is a tough job, and it requires practice. Computer experts have found it surprisingly difficult to make a robot that can play sports or even walk steadily. The instructions are complicated.

Physical activity in children is correlated with a number of positive characteristics. Physical education programs that emphasize enjoyment over competition or criticism are most successful at producing all these benefits. Active children have higher self-esteem than inactive children. Feelings of stress, depression, and anxiety are less common in active than in inactive children. Some of these results may occur because depressed children are reluctant to exercise. But on the other hand, a few small intervention studies show that exercise modestly reduces anxiety and depression symptoms in children, as it clearly does in adults. Although parents sometimes worry that physical activity will take up time better devoted to academic effort, no study has shown a drop in academic performance as a consequence of increased activity. A few studies have even shown that fitness is associated with higher achievement test scores in children.

Indeed, a considerable body of research shows that exercise is vital for cognitive ability. Across the lifespan, heart health is brain health. In childhood, aerobic fitness is correlated with math and reading achievement, while muscle strength and flexibility are not. In a meta-analysis, children (ages four to eighteen) who were more physically active tended to score higher on tests of IQ, perceptual skills, verbal ability, mathematical ability, and academic readiness (with an effect size, d', of 0.32; see figure, p. 64). The relationship was strongest from ages four to seven and from eleven to thirteen. Fit children perform better on attention and conflict cue tasks (see p. 116), both of which are associated with self-control ability. Fitness also improves performance on a *relational memory task*, remembering which pairs of pictures were shown together, which requires both the prefrontal cortex and the hippocampus.

Parents should take advantage of these active years to introduce their children to sports that can become lifelong hobbies.

Though the habit is best started young, the brain benefits of regular aerobic exercise are especially noticeable in old age. Problems with an aging circulatory system can reduce the blood supply that brings oxygen and glucose to your brain. Elderly people who have been athletic all their lives are better at planning and organizing their behavior and dealing with ambiguity or conflicting instructions than sedentary people of the same age. When inactive people get more exercise, even in their seventies, their brain function improves in just a few months. Exercise is also strongly associated with reduced risk of dementia, including Alzheimer's disease, late in life.

Physical activity leads to growth and increased activity in the frontal and parietal cortex in adults. In brain imaging studies, these areas have been observed to be active during tests of reading comprehension and mathematical calculation, as well as during many tests of self-control. Fit adults, young or old, show reduced activity (suggesting increased efficiency) in the anterior cingulate cortex, which is involved in detecting errors and resolving conflict between alternatives.

Brain differences related to exercise have also been observed in children. In physically fit nine- and ten-year-olds, the hippocampus and the ***dorsal** striatum* (a region important for response selection, cognitive flexibility, and the performance of learned behaviors) are larger than in sedentary children. These differences are related to improved performance on behavioral tasks that rely on these areas. The role of exercise in maturation of the prefrontal cortex of children has not yet been examined, but the behavioral effects of exercise suggest that this brain region might be affected as well.

There are several possible ways that exercise might change the brain, and they are not mutually exclusive. Neuroscientists found that in people and lab animals, exercise causes the release of *growth factors*, proteins that support the growth of dendrites and synapses and increase synaptic plasticity. Exercise also increases the survival of newly formed brain cells in the hippocampus of young

PRACTICAL TIP: PROTECT YOUR CHILD FROM HEAD INJURIES

There's one big exception to the rule that exercise is good for your brain: the risk of head injuries from contact sports. Consider the case of a twenty-one-year-old college (and high school) football player named Owen Thomas, who killed himself in 2010. He is the youngest person known to have had chronic traumatic encephalopathy, a neurodegenerative disease that was found at autopsy after his suicide. (**Neurodegeneration** is a general term referring to progressive loss of structure and function of neurons.) He had never been diagnosed with a concussion and had no previous history of depression, but he was known as a hard hitter on the field.

Multiple studies have shown that repeated blows to the head increase the likelihood of developing neurological symptoms later in life, including dementia, movement disorders, and depression. As 1.6 to 3.8 million sports-related concussions occur each year in the U.S. alone, this is potentially a big problem. Girls are more likely to be diagnosed with concussions than boys in the same sports, although the difference may be simply that boys are less likely to report symptoms.

Chronic traumatic encephalopathy has a variety of symptoms, including depression, memory loss, and impulse control problems. Its symptoms are easily confused with those of depression, Alzheimer's disease, Lou Gehrig's disease, or Parkinson's disease, so its prevalence remains uncertain. (Some researchers now believe that Gehrig himself may have had chronic traumatic encephalopathy, misdiagnosed as the disease that was named for him.) Individuals who carry the ApoE4 gene variant are at higher risk of Alzheimer's disease, and this variant also seems to increase the risk of neurodegeneration associated with head injury.

Many popular childhood sports may be risky. Football, wrestling, rugby, hockey, karate, lacrosse, soccer, basketball, and skiing are all associated with repetitive head injury. It is not yet clear how the number, timing, and severity of head injuries relate to these long-term risks. Experiencing three concussions significantly increases risk in most studies, and some studies find increased risk after a single concussion (or undiagnosed head injuries, as in Thomas's case).

Compared to adults, children take longer to recover from a concussion. This may be because there is less space between the brain and the skull in childhood, making brain swelling after injury a more serious problem for children than adults. Children are also especially susceptible to complications from reinjury during the recovery period. Over 90 percent of deaths due to sports-related head injury since 1945 in the U.S. have been athletes who had not yet finished high school.

Though longitudinal research is needed to sort out these questions, caution seems well justified already. Children whose family history puts them at risk of Alzheimer's disease should be especially careful to avoid head injury. No child should return to play with any lingering symptoms of a concussion. (Not all concussions lead to loss of consciousness; confusion and memory loss after a head injury are also signs of concussion.) During recovery, children should rest their brains as well as their bodies, avoiding schoolwork and even reading. Finally, parents should seriously consider a nonimpact sport for children who have suffered more than one head injury in their current sport. For more information, see http://www.cdc.gov/concussion.

adult and old lab animals—and even in rat pups whose mothers exercised during the pregnancy. Proteins that are released during exercise in juvenile and old lab animals stimulate blood vessel growth, helping nutrients from the circulation to reach the brain. Most of these effects have been studied in adult animals, but they probably also occur in children's brains. The question of how exercise interacts with brain development remains to be studied.

To gain the benefits of exercise on the brain, children of all ages should have fun moving their bodies for at least an hour a day. Young children rarely need much encouragement to run around, so most kids meet or exceed the activity requirement through elementary school. In addition to letting kids play tag and climb trees, which are unlikely to continue into adulthood, parents should take advantage of these active years to introduce their children to sports that can become lifelong hobbies and social activities, such as martial arts, dance, softball, or hiking. Parents can also help by limiting children's exposure to sedentary activities like TV and computer time (see chapter 16) and by demonstrating to their children that exercise is a regular feature of adult life.

In our fat-obsessed culture, it is tempting to focus on weight control as a key benefit of exercise, but this approach is likely to be counterproductive (see *Practical tip: Worried about your child's weight?*, p. 108). Thinking of exercise as a part of dieting defines it as an occasional activity, not something you do every day for fun—and makes it feel like a punishment for bad behavior, especially to young girls. No one looks forward to activities associated with shame and guilt, so using this type of motivation is likely to be actively harmful to your attempts to make exercise a regular and enjoyable part of your child's life. Exercise is beneficial for everyone, fat or thin, whether or not it leads to weight loss.

There are also social reasons for children to master motor skills when they're young. A swing and a miss, which is cute in a preschooler playing T-ball, may be mortifying to an adolescent boy in gym class. The embarrassment associated with attempts to catch up with the better-practiced skills of peers often leads to a lifelong distaste for athletics. In longitudinal studies, an inactive childhood is a strong predictor of an inactive adult life.

Parents' efforts in laying this sort of groundwork during middle childhood are likely to pay off in adolescence and adulthood. Physical activity typically decreases in the early teen years, on average at age thirteen for girls and fifteen for boys, as organized activities and socializing squeeze out competitors like exercise and sleep (see chapter 9). For many kids, this stage of development marks the divide between an active childhood and a sedentary adulthood, with substantial consequences for adult health.

Through most of our evolutionary history, people were much more active than they are today. Our bodies and our brains are adapted to regular exercise, and they do not function well without it. There are exercises for every child's taste, from solitary distance running to competitive team sports to sociable yoga classes. Helping your children find a few that suit their personalities and talents will go a long way toward giving them a good start in life.

Chapter 16

ELECTRONIC ENTERTAINMENT AND THE MULTI-TASKING MYTH

AGES: BIRTH TO EIGHTEEN YEARS

Both of us are young enough to be distracted frequently by e-mail or the Web when we're trying to work, but we're old enough that our brains didn't develop under those conditions. In contrast, many of today's children are growing up with continuous access to electronic media, from the TV in the bedroom to video games for the road. In the U.S., the average baby starts watching TV at five months of age, before he can sit up by himself. By seventh grade, 82 percent of children are online.

Your child's brain is wired to seek out and pay attention to new information because our ancestors' survival often depended on detecting changes in the environment—from the arrival of a lion to the new expression on a mate's face. But what happens to our brains when getting new information becomes too easy? Our society seems to be in the process of finding out, as the Internet brings an avalanche of facts and ideas (along with daredevil stunts and keyboard-playing cats) into our lives.

Researchers do not yet know the full effects of exposing children to a constant stream of highly salient stimuli, as the new media environment is a recent occurrence. Let's start with some basic conceptual principles, established by decades of neuroscience research. Throughout this book, we have emphasized that your child's brain depends on commonly available sensory experiences to help determine which neural connections should be maintained or lost. If the character of these experiences changes dramatically, we would expect to see effects on brain development. In the case of new media, the few experiments that have been done support this idea, though not with certainty. One thing we do know is that the effects of media experience on the brain depend on the details—whether your child is passively watching TV or actively playing video games, at what age the experience occurs, as well as which activities your child is neglecting in order to find time for it all.

At any given moment in everyday life, deciding which information should have your attention is far from simple. Your brain needs to combine the ability to concentrate deeply on a particular problem with the ability to reorient itself quickly if important changes happen. To achieve this balance, top-down attention, the deliberate direction of your brain's focus by the cortex, must compete with bottom-up attention, the automatic capture of your brain's focus by salient events, such as someone screaming—or the buzz of your cell phone reporting an incoming text message. You've probably experienced moments when one system dominated the other: either your concentration was so deep that you didn't hear your baby cry, or you were getting interrupted so often that you couldn't make any progress on your work assignment.

The bottom-up system is functional at birth, earlier than the top-down system, which can take control for brief periods by the end of the first year and continues to improve at least to age ten and perhaps into adolescence. This discrepancy in the maturity of the two systems is why younger children are easier to distract than older children. As we discussed in chapter 13, the ability to use executive function is a key ingredient of self-control and improves with practice.

A basic principle of neuroscience is that brains become better at doing whatever they do frequently. Video game players are an excellent illustration of this point, as they typically invest hundreds or thousands of hours into practicing games that require the rapid detection of targets and their discrimination from nontarget distractors. Daphne Bavelier and her colleagues showed that this effort improves response speed (across a variety of tasks) and visual attention abilities of players' brains. Although you might imagine that video games would train bottom-up attention, in fact the benefits seem to be mainly improvements in the effectiveness of the top-down attention system. Unfortunately, only shoot-'em-up rapid action games, of the type that parents hate, seem to have these effects.

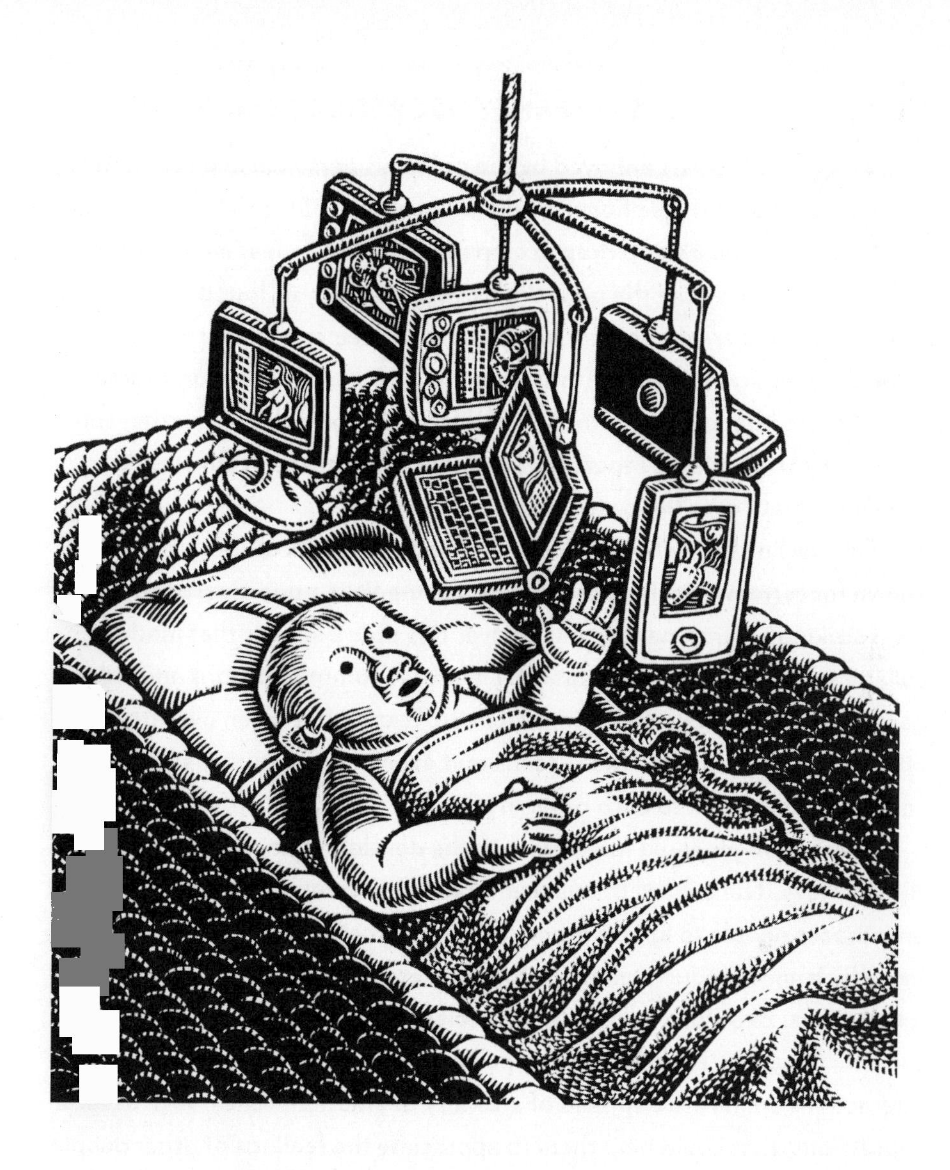

The changes appear to be a direct effect of video game playing. Alternately, you might imagine that people with better visual attention would find video games more rewarding and so be more likely to play them, which would be an example of reverse causation. But researchers ruled out that interpretation because, in some studies, college students who did not play regularly were randomly assigned to practice action video games (for example, *Grand Theft Auto* or *Call of Duty*). The control group spent an equal amount of time practicing nonaction video games (*Tetris*, *The Sims*). The training improved several aspects of visual attention, but only in the action video group. This group also improved in contrast

SPECULATION: DOES INTERNET USE REDUCE EMPATHY?

Who isn't annoyed by the oblivious person in line who won't get off his phone to talk to the cashier? Since 2000, there has been a sharp drop in empathy in the U.S., leading to experiences like the poor clerk's. A meta-analysis found that today's college students are 40 percent less likely than their counterparts two or three decades ago to agree with statements like "Before criticizing somebody, I try to imagine how I would feel if I were in their place." In other surveys, many U.S. citizens report feeling that people have become less kind and considerate over the past decade.

This decline has only been identified recently, so its causes are not known for certain, but the rise of electronic media is a prime suspect. The most striking environmental change over this time period is that modern college students have grown up with electronic communications and video games as a constant presence in their lives. Texting has taken over from phone conversations, and many teens text even in situations where they could turn and speak to each other in person.

These changes could interfere with the development of empathy in a variety of ways. The culture of online interaction in communities like Facebook encourages people to ignore others who are perceived as overly demanding, which might reduce tolerance for people with problems in face-to-face interactions. Exposure to video game violence might desensitize children to the pain of others. Perhaps most important, children who learn about social interactions online are deprived of a variety of emotional cues, such as facial expressions, that could help them to appreciate the feelings of other people. Without such real-world experience, empathy may not develop correctly—and that could be very costly for all of us.

sensitivity (the ability to see faint outlines, like cars in fog), an effect that lasts for years after training. Players react more quickly than nonplayers, with no difference in the number of errors they make in their reactions.

Similar randomized experiments cannot be done in children because it would be unethical to expose them to the violence in action games. We would guess that if there is a difference between children and college students, it is likely to be that young brains are more plastic than adult brains. Researchers are looking for nonviolent games with similar beneficial effects, but they have not yet managed to find any.

Children who are already action video game players at home show most of the same advantages over nonplayers (or players of nonaction games) that are found in college students. Across all ages from seven to twenty-two years, players are better able to orient their attention to a cued location, compared to nonplayers of the same age (in the conflict cue task, see p. 116). Distractor arrows (in the incongruent condition) slow the responses of players proportionately more than those of nonplayers, suggesting that video game players pay attention to a broader field of view. Players are also better than nonplayers at detecting a target against a background of distractors, at rapidly processing a stream of visual inputs, and at tracking multiple targets simultaneously. Because video game players are mostly boys, these effects of practice may lead to gender differences in attentional processing, which have not existed in previous generations.

Before age two, electronic entertainment has no benefit and clear costs.

Another recent change in children's sensory experiences is the rise of multitasking. In the U.S., children now fit 8.5 hours of media use into 6.5 hours per day, on average, by doing multiple things at once (text messaging while watching TV or listening to music while playing video games). Almost all of us believe that we can multitask. However, the brain cannot concentrate well on more than one thing at a time. A major source of interference between tasks seems to be the limited capacity of the prefrontal cortex, a key aspect of the brain's executive control system.

Many aspects of brain function, such as walking or driving a familiar route, do not require direct conscious control. Planning your actions, though, does require attention, as you'll notice the next time you get distracted and end up at a familiar destination when you meant to go somewhere else. People who claim to do multiple attention-demanding tasks at once are actually switching between the tasks repeatedly. Every transfer of attention from one task to another requires

PRACTICAL TIP: BABY VIDEOS DO MORE HARM THAN GOOD

Almost all children today start watching TV before they are two years old. This is not good news for their future. Marketers claim that baby videos, such as those available from Baby Einstein and Brainy Baby, can give your child's brain a head start, but research shows that the opposite is true.

As we have discussed, sensory experience is important for brain development, particularly the growth and retraction of synaptic connections, which is vigorous during the first three years of life. We would expect that children's brains should be affected by watching TV for as much as a third of their waking hours during this developmental window. The key question is *how* are they affected?

Babies' brains are optimized to learn from social interaction. For example, babies learn poorly from exposure to a second language on video (see chapter 6). The electronic babysitter reduces the time that infants spend interacting with other people, which can impair many aspects of their development. Even if you watch TV with your baby in the room, without intending for her to watch it, its presence interferes with her play and reduces the amount of social interaction between you.

Multiple studies have shown that infant TV watching is correlated with poor language development. U.S. babies of seven to sixteen months who spend more time in front of the screen know fewer words. Two or more hours per day of screen time before the first birthday is associated with a sixfold increase in the risk of language delay in Thai children. Even *Sesame Street* viewing by babies correlates with language delay, though this program has lasting beneficial effects on three- to five-year-olds. Exposure to baby videos is also associated with reduced cognitive ability at age three.

The quick cuts and bright colors that are typical of baby videos may also interfere with the normal development of attention. Before ten months of age, your infant cannot direct his attention voluntarily (see chapter 13). Exposure to attention-grabbing stimuli like fast-paced entertainment programs may make the transition to voluntary attention more difficult. In a longitudinal study, children who watched violent shows before age three were more than twice as likely to have ADHD at ages five to eight. It is easy

to imagine that parents of ADHD children might be tempted to use TV to occupy their child, creating a vicious cycle (see chapter 28).

No reliable research shows that TV watching has any benefit for babies. France recently banned programming directed at infants, but it is unlikely that the U.S. will follow suit. Instead, parents need to protect their babies by keeping them away from the TV until they are at least two years old.

resources, as your brain must remember or reconstruct where you were on the abandoned task when you come back to it. The first task can also interfere with the representation of the second task in your memory. For these reasons, switching back and forth between tasks takes more time than completing the same tasks one by one in sequence. What cognitive psychologists call *costs of switching* reduce performance on individual tasks. If both tasks are highly practiced, this switching cost can be reduced or perhaps in rare cases eliminated. Under no circumstances, though, is it more efficient to do multiple attention-demanding tasks at once than to do them separately. In short, multitasking is a myth.

The cost of chronic multitasking may include diminished performance when single-tasking. One study found that college students who multitask more often performed worse than those who multitask less often at tests of distractibility and task-switching ability. Students who spent more hours using the Internet also tended to spend more time using multiple types of media simultaneously. The performance of high multitaskers was more severely impaired by distractors, compared to low multitaskers, though the two groups performed equally well when no distractors were present. In other tasks, high multitaskers also had more trouble ignoring irrelevant items in their memory and switching back and forth between two sets of rules. One thing we don't know is whether naturally distractible people are more likely to multitask or whether multitasking actually increases distractibility. Either way, though, if your children spend a lot of time multitasking, it's not likely to be good news for their ability to concentrate.

As we've discussed throughout this book, your children's life experiences influence their brains in a variety of ways. Because the rise of electronic media is both so recent and so significant, scientists are working hard to determine its effects on child development. We do know that young children's brains are strongly influenced by one-on-one interactions with caring adults, which cannot be replaced by anything that appears on a video screen. Before age two, electronic entertainment has no benefit and clear costs (see *Practical tip: Baby videos do more harm than good*, p. 142). For older preschoolers, educational TV can be beneficial, while certain programs are more likely to do harm. For example, children who watched *Dora the Explorer* or *Blue's Clues* at age two and a half had better language skills, while those who watched *Teletubbies* had worse language skills than average. In school-age children, electronic media are a mixed blessing, improving some cognitive capacities, while perhaps impairing others.

Banning electronic media throughout childhood is probably unnecessary and certainly unrealistic. Even if you could enforce such a ban successfully, cutting your child off from his peers' favored communication methods would place him at a severe social disadvantage, and lack of experience with computers might lead to professional difficulties later on.

Even so, there are strategies you can use to reduce the negative impacts. Providing time for other activities by limiting screen time is likely to be beneficial for development. There is no doubt that childhood experiences influence brain development, so it's worth making sure that the important ones are available to your child. Children need exercise (see chapter 15), they need face-to-face interaction (see chapter 20), and they need to spend time outdoors (see chapter 10). In that light, your mom's suggestion to "Go outside and play" is backed not only by common sense but by modern brain research. Dandelions need to see the sun every so often.

PART FIVE

YOUR CHILD AS AN INDIVIDUAL

NICE TO MEET YOU: TEMPERAMENT

EMOTIONS IN THE DRIVER'S SEAT

EMPATHY AND THEORY OF MIND

PLAYING NICELY WITH OTHERS

Chapter 17

NICE TO MEET YOU: TEMPERAMENT

AGES: BIRTH TO TWENTIES

Many people start reading parenting advice books before their baby is born—not to learn how to change a diaper or breast-feed, but to find out how to make their baby intelligent or sociable. If it doesn't seem a little odd to be trying to figure out how to build a good relationship with someone you haven't met yet, perhaps that's because your culture assumes that babies are blank slates who can grow up to become any kind of person, no matter what their individual characteristics may be. You can see this kind of reasoning across the political spectrum: from liberal parents who are horrified to find their son playing war with sticks after being forbidden to have toy guns to conservative parents who believe that therapy can turn their gay son into a straight man.

As we have said, most children flourish to a similar extent regardless of circumstances, as long as conditions are "good enough." This is true of personality, where, as with so many other brain functions, the vast majority of children are dandelions. Indeed, there are limits on how much parents can influence their children's personalities. It may sometimes seem otherwise, but this is good news: you do not bear the sole responsibility for your children's success and happiness in the world.

Although your child's personality is indeed shaped in part by environmental influences, it's worth remembering that not all of them come from you. This may be easier to understand if you keep in mind that the nuclear family is a recent

invention in the history of our species. During most of our evolution, children were raised by a community and were often cared for by older children as well as adults. For many cultures, the situation is still the same today. Wherever you live, your child is likely to spend much of his time with others: in school, at play, and with siblings, friends, teachers, and other people. Taken together, these factors are usually sufficient to lead to a good outcome (though see chapter 30 for what happens when environmental conditions are not good enough).

First-time parents sometimes manage to hold on to the blank-slate illusion for a while, but anyone who's had more than one child has been clued in that this idea is wrong. Even at birth, babies differ from each other in important ways. *Temperament* is a psychological term for the individual differences in infants that form the basis for adult personality. Temperamental characteristics include activity level, attentional persistence, and how easily the child feels fear, anger, frustration, or happiness. Like most early differences between individuals, temperamental characteristics do not absolutely determine outcome, but they do influence the relative likelihood of various adult personality traits and the range of possibilities available to a particular child. Perhaps more important for parents, they also influence your child's response to different parenting styles—and vulnerability to various developmental problems.

Even at birth, babies differ from each other in important ways.

One of the best-studied aspects of temperament is whether babies react calmly or with distress to unexpected events. Nearly one in five infants are *high-reactive babies*, who kick and cry when exposed to something new or strange (but not especially scary), like a dangling mobile over the crib. As they grow into toddlers, such children tend to be what the scientific literature calls *behaviorally inhibited*—and what the rest of us would call shy or reserved. When high-reactive babies are followed into early adulthood, they are likely to be introverted and are at higher risk than calm babies of growing up to be worry-prone adults, even if they don't have a formally diagnosed anxiety disorder.

Various experiences can nudge a high-reactive child's developmental path in a less-anxious direction. For example, infants who are unusually irritable at fifteen days of age are more likely than other babies to become insecurely attached to their mothers at one year (see chapter 20), but this outcome is much less likely if the mothers are trained to soothe restless babies effectively. In general, about half of all high-reactive babies raised in the U.S. become fearful children sometime before age seven. The good news is that only a third of these babies become anxious adults. The others tend to become detail-oriented, well-prepared, and thus often successful people. Among other things, they make good scientists (Sandra was a high-reactive baby), though it's rare for them to be the life of the party or to pursue careers as politicians or salespeople.

Among adults without anxiety problems, those who started life as high-reactive babies are more vigilant than people who were calm as infants. In one laboratory test, vigilant adults find it difficult to disengage their attention from threatening faces on a computer screen. Most people become tense when watching a blue screen that signals that they might be hit with an unpleasant air puff, but only adults who were high-reactive babies stay tense in the presence of the green screen that signifies safety. These laboratory reactions don't matter much in everyday life, but there is a broader, important point: there are real biological differences underlying high-reactive temperament, and these differences persist into adulthood.

Among children or adults, personality doesn't require that you must act a certain way all the time. Instead it sets thresholds that make you more or less likely than average to behave in certain ways in a given context. For instance, most shy people don't feel inhibited in front of family members, and even an outgoing person might feel shy while speaking to an audience of thousands. Still, personality has an important influence on life outcomes. Indeed, personality traits are as effective as IQ (see chapter 22) or socioeconomic status (see chapter 30) at predicting the probabilities of various life events, such as divorce or work success.

The most widely accepted model of adult personality contains five factors: openness to experience, conscientiousness, extroversion, agreeableness, and neuroticism. These factors vary among individuals across many cultures, from Malaysia to Estonia. The stability of personality traits increases with age, being relatively stable after thirty years of age, especially from fifty to seventy. All five factors show strong heritability, meaning that people who are more closely related are more likely to share personality traits. For this reason, you are likely to have some personality characteristics in common with your biological parents, even if you are adopted and have never met them.

On the other hand, even identical twins have noticeably different personalities, so some life events must help to determine how personality matures. It has proven surprisingly difficult for researchers to determine which aspects of experience influence how temperament unfolds into adult personality. Behavioral genetics studies consistently find that the environmental influences shared among children in the same family have little or no effect on their adult personalities. Indeed, identical twins reared together are no more similar in their personalities than those reared apart. As you may recall from chapter 4, though, these studies sweep up gene-environment interactions into the "genetic" category, which creates some

DID YOU KNOW? WHY YOU'RE TURNING INTO YOUR MOTHER

On average, when it comes to personality, children turn out much like their biological parents. One reason is that your genes affect your life experiences. Babies' temperaments influence how other people interact with them—a lot. People tend to talk about kids who grow up in the same family as if they share the same environment, but there are many important differences, including the ways that each child relates to parents, siblings, and eventually peers.

From the moment a baby is born, parents react to his or her individual characteristics. Parents would find it impossible (and senseless) to speak to a willful toddler in the same way that they do to an amiable one. Similarly, a child who is sociable will get more practice at speaking and listening than one who's most interested in playing with his train set in another room. These are among the many reasons why it makes no sense to talk about heredity and environment as separable factors in development.

As children grow, their innate differences in temperament and the resulting individual differences in experience often reinforce each other's effects. Older children can control their environments to a greater extent than younger children, for example, by choosing to take gymnastics lessons instead of reading books in their free time. Perhaps for this reason, individual differences tend to become more pronounced later in life—and most obvious of all in adults. Genetically related people share more personality traits in adulthood than they do in childhood. People's increasing ability to choose environments that are well matched to their individual tendencies could explain why.

misperceptions. This is important when it comes to personality, where gene-environment interactions have important effects.

Although parents don't like to admit it, individual children within the same family are raised in different ways, often linked to their temperaments (see *Did you know? Why you're turning into your mother*). In addition, the same environment may have different effects on children with different genes. As researchers have started to look for examples of how gene-environment interactions influence personality development, they have found a variety of effects. Temperamentally anxious children develop more empathy in response to gentle child-rearing techniques than do bolder children (see *Practical tip: Promoting conscience*, p. 177). Children with a specific receptor that makes them prone to hyperactivity and impulsivity (including ADHD; see chapter 28) are more sensitive to parenting style than other children. Eventually, parents may be able to personalize their interaction styles to produce a desired outcome based on the characteristics of an individual child, but researchers have a lot of work to do before that dream becomes reality.

The same environment may have different effects on children with different genes.

The development of antisocial behavior is one well-studied example of a feedback loop that starts with child temperament, which affects parental behavior, which then further modifies the child's behavior. Children who are irritable or prone to aggression are challenging to raise, making them less dandelion-like than most children. Both biological and adoptive parents often respond to the child's frustrating behavior with harsh restrictions and punishments. Parents who have their own problems with aggression are more likely to produce children of this type—and also more likely to discipline them harshly. In addition, parents who are frustrated for other reasons, such as a troubled marriage or job insecurity, are more likely to respond harshly to their children. This rough treatment in turn increases the child's aggressive behavior until eventually it may become uncontrollable. Interventions that reduce parental harshness also reduce the risk of future aggressive behavior in the child.

Environmental events that influence child development are not restricted to the family. All children have a life outside the house, and many of their interactions with the world leave permanent traces behind. They spend much of their time with teachers and friends, taking part in sports or other activities. Children learn a lot from their peers. For example, children of immigrants typically speak with the accent of their friends, not their parents, and they learn to speak the language of their peers fluently even if their parents don't speak it at all. Children's attitudes and behaviors typically change over time to become more similar to those of their peer group, and this influence can be positive. For example, a low-achieving child who

MYTH: BIRTH ORDER INFLUENCES PERSONALITY

Firstborn children are self-reliant, traditional, and successful, while last-born children are easygoing, creative, and rebellious? Actually, no. Despite the bottles of ink that have been spilled defending this idea, siblings show no consistent personality differences based on their place in the family. Thousands of psychology papers have been published on this topic—most of them flawed.

One of the biggest flaws is a demographic one. Both small and large families contain firstborn children, while children born third, fourth, fifth, and so on are, by definition, only found in larger families. Many studies failed to control for differences in socioeconomic status between small families (generally well-off and educated) and large families (typically poorer and less educated). So firstborn children, on average, have advantages over later-born children, simply because a larger proportion of the firstborns come from a small family. Many studies that claim to show greater success in firstborn children than in later-born children suffer from this conceptual error.

A second problem arises when researchers ask parents to rate their children's personalities. Generally, these ratings do not agree with ratings by outside observers. Birth-order effects on personality are perceptible when people are evaluated in the context of their families, but almost nonexistent in the outside world. Part of the difficulty is that parents are necessarily comparing an older child with a younger child, and age is one of the strongest predictors of maturity for any personality type. Parent-rated studies are more likely than those that evaluate personality in other ways to find that firstborn children are more mature than later-born ones. Another concern with this approach is that people act differently within the family than they do outside it, as is clear to any adult who goes home for the holidays and feels instantly reduced to the age of twelve.

In a meta-analysis that only included studies with controls for family size or socioeconomic status, the remaining effects were small and inconsistent. More than half the studies found no effects of birth order on personality at all. And those that did show a pattern were more likely to be small studies, with few subjects, where chance plays a bigger role. This is the opposite of

what we would expect for a real effect. By chance, small studies are more likely to be flukes, while large studies have more statistical power and are thus more reliable. The largest study of this topic, with over seven thousand subjects, found no differences in personality between first- and second-born children in families of two children. Sorry, firstborn readers, but there is little credible evidence that birth order influences personality.

falls in with a group of high-achieving friends is likely to improve his schoolwork. Of course, children choose their friends, rather than being randomly assigned to peer groups (as a researcher might prefer), so much of the similarity between children and their friends results from the selection of friends who are already similar in attitudes and interests. In longitudinal studies, though, children do become more similar to their friends than they were when the friendship started.

Culture strongly influences how temperament develops into adult personality—another example of how brain development matches children's behavior to their environment. Behavioral inhibition is accepted and encouraged by Chinese mothers, who interpret it as reflecting self-restraint and maturity. In contrast, Canadian mothers endeavor to draw out children who show behavioral inhibition, which in that culture is believed to reflect fearfulness and lack of social skills. Accordingly, high-reactive children in China are more likely than those raised in Western cultures to grow up to be reserved, a trait that leads to social success—in China. In general, the response of a parent to a child's temperament is more influential if it's consistent with the beliefs of the culture they live in.

Perhaps personality development is too complicated to study in people, since scientists can't control (or measure—or perhaps even identify) all of the influences that might matter over people's long childhood. Indeed, the clearest evidence that parenting influences personality comes from animal studies. Rat mothers who lick and groom their babies a lot produce offspring that are less timid and more prone to exploration. This is true even if the rat pups are born to low-grooming mothers but raised by high-grooming mothers, in the equivalent of human adoption studies. Similar studies show that high-reactive monkeys are more vulnerable than low-reactive ones to variations in the quality of mothering, and as we noted above, some evidence suggests that the same is true of children. (For more details on this research, see chapter 26.)

Even if parents cannot entirely control how their children turn out, parenting still matters. First and most important, your relationship with your children is its own reward. How you get along, both as they're growing up and after they're adults, depends on how well you care for them. Second, your children's behavior at home depends a lot on your household rules and how you enforce them (see chapter 29). This can have a strong effect on the happiness of their (and your) home life. Third, you can teach your children a wide variety of skills that are useful in adulthood, from cooking to financial literacy to car repair. You can also give them opportunities to discover their passions. Fourth, you can help your child learn strategies to live comfortably and productively with his or her individual temperament, especially if it's one that you both share.

Think of parenting not as growing the person you want, but as a process of helping your child discover how to make his or her unique abilities and preferences fit well with the rest of the world.

Chapter 18

EMOTIONS IN THE DRIVER'S SEAT

AGES: BIRTH TO EARLY TWENTIES

All of us have experienced emotions that seemed overwhelming and out of control. Imagine feeling that way much of the time, and you have a picture of your young child's daily experience. One reason that life with toddlers is such a wild ride is that the parts of the nervous system that produce raw emotions mature earlier than the brain regions that interpret and manage them. It's probably just as well that your children won't remember that stage of their life. If only we could provide the same service to parents when it's one of those days.

Emotions organize our minds. As basic survival signals, emotions are present from birth, though they become increasingly more complex as children grow up. At the most fundamental level, emotions (unlike moods) are reactions to the environment that help you to rapidly distinguish between rewarding and threatening aspects of the world. Emotions also compel you to pay attention to salient events, define what you value in life, prepare your body for action, and communicate your internal states to other people.

Certain emotions are universal, occurring in all cultures that have been studied. The facial expressions that signify these so-called basic emotions (fear, joy, disgust, surprise, sadness, and interest) are also built into human biology, so that you could understand whether someone was glad to see you or angry, even if the two of you shared no history, language, or cultural heritage.

Newborn babies can smile, but they do not begin to smile in response to external events, such as faces, voices, or bouncing, until they are three to eight weeks old. By about three months, they smile more at familiar faces and show interest and surprise. By four months, babies are adept at laughing, and their appreciation for visual games, like peekaboo or seeing funny faces, continues to climb throughout the first year.

Signs of negative emotion are also present early in life. The earliest to appear are startle, disgust, and distress, which (as with smiling) may not be related to external events in the first two or three months. Anger and sadness show up in facial expressions at three months, usually provoked by pain or frustration. All these emotional expressions help to evoke care from parents and other adults.

You may remember the first time your baby smiled back at you. His ability to recognize emotions in other people's faces develops almost as early as his ability to show facial expressions. By two or three months of age, the occipitotemporal cortical region, which is specialized for face processing, is already activated more by faces than by other objects, though its tuning at this stage is considerably broader than it will be in adulthood. At seven months, infants stare longer at a fearful face than they do at a happy or a neutral face, and their frontal cortex shows an electrical response associated with salient stimuli.

The amygdala sits at the center of the brain's network for processing emotions (see figure opposite). It receives input from a wide range of brain regions, including information from all your senses. The amygdala's outputs also go to many brain areas, which constitute two major systems. One, acting via the hypothalamus and brainstem, activates the autonomic nervous system to produce changes in heart rate, blood pressure, and breathing that prepare the body for fight or flight (see chapter 26). The other, via various regions of the cortex, controls the cognitive aspects of emotion, including interpretation, regulation, conscious perception, and emotional reactions to memory and imagination. These connections are reciprocal, with both systems influencing the amygdala in return.

Because these connections are so widespread, emotions influence almost every system in the brain. Laboratory studies have shown that emotional signals can improve visual perception, and their fingerprints are all over decision making. Even something as simple as choosing between a blue shirt and a green shirt is difficult for patients with damage to the emotional regions of the cortex.

The amygdala prioritizes speed over accuracy, so it sends out a lot of false alarms. For instance, if you are walking in the woods and see a curved stick on the ground, you might jump back quickly, fearing a snake, before you have time to realize your error. Such a response is encouraged by a hardwired legacy from our evolutionary history. Often identified with fear, the amygdala actually has a broader mandate: it assigns value to stimuli, priming your brain to react appropriately based on your previous experience with that situation, person, or object. These values can be positive as well as negative, but they're not very sophisticated. If your visual system isn't sure whether that dark spot is a spider or a piece of dirt, the amygdala assumes it's a spider until the cortex corrects that impression.

Even as infants, children show a wide range of individual variability in their tendency to express positive and negative emotions, which is a component of temperament (see chapter 17). Some of these differences result from genetics; identical twin infants are more similar than fraternal twins in sociability,

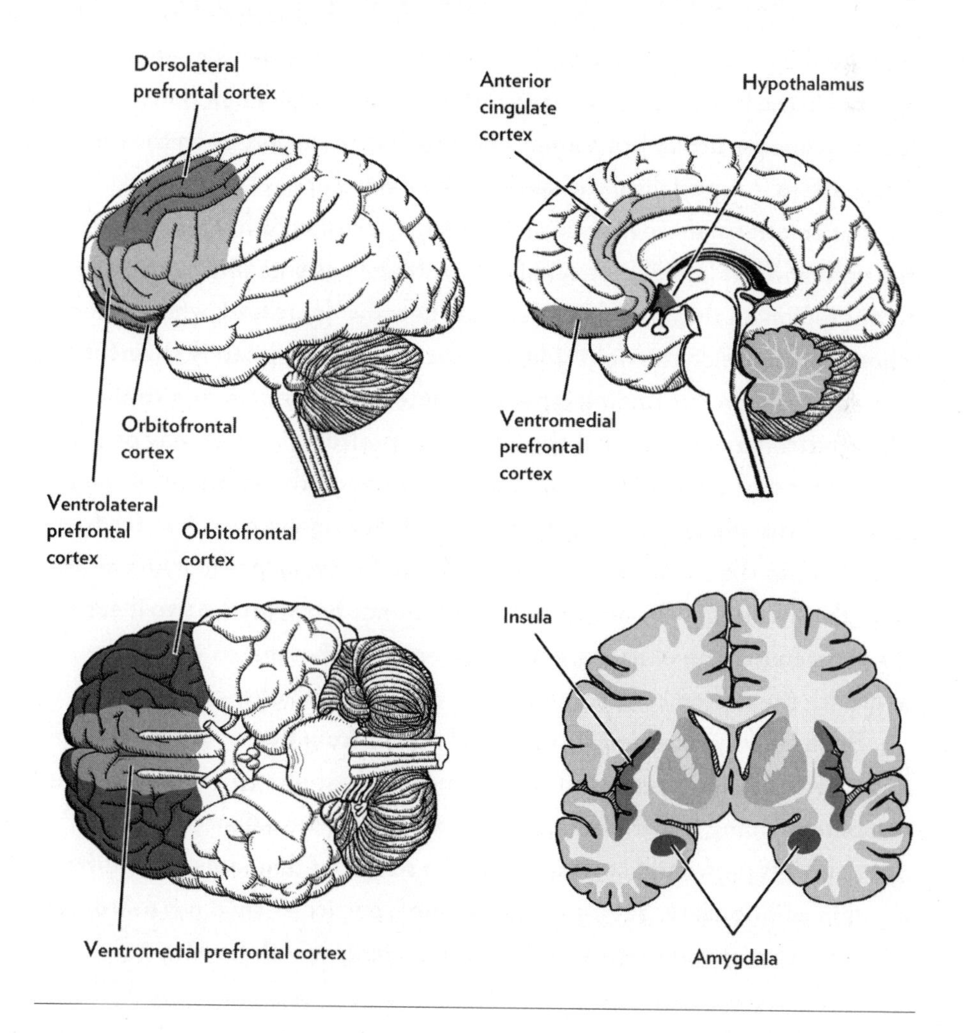

MYTH: THE RIGHT HEMISPHERE IS THE EMOTIONAL SIDE

You've probably heard people talk about being *left brained* when they mean logical and *right brained* when they mean emotional. This hypothesis was formulated several decades ago, before the invention of functional brain imaging. It is true that the emotional content of speech (its tone or prosody) is processed by language areas on the right side of the brain, but in general emotions are equally effective in activating regions on both sides of the brain. And the two halves of the brain are so heavily interconnected that it makes no sense to claim that an entire half is somehow left out of the mix.

The basic principle is not that the left/right brain is rational/emotional, but that specific functions are localized. Evolution has selected for brains that use the least axonal "wiring," so related functions often sit near one another in the brain. In many cases, different aspects of a function get collected in one particular brain area, which can be on the left or on the right. This pattern is especially noticeable in big-brained apes like us.

For emotions, there may be an interesting division of function between the right and left lateral prefrontal cortex. Many patients who have damage to this region on the left side of the brain become depressed, while patients whose damage is on the right side may become inappropriately cheerful or manic (though this finding is not completely consistent across studies). These outcomes suggest that the left lateral prefrontal cortex may be specialized for positive emotions and the right for negative emotions. A related version of this idea holds that the left side carries signals for approach (getting closer to the situation), and the right side for avoidance (staying away from the situation). The only practical difference between the two ideas is the characterization of anger, a negative emotion that makes you want to approach the person causing it (in order to punch his lights out).

Brain activity recorded via electrical signals from the scalp supports this hypothesis. The difference between hemispheres shows up very early in development. Some researchers have suggested that the balance between the right and left prefrontal cortex may be the source of temperamental differences in emotionality, determining whether a particular child has a stronger tendency to respond to the world by approaching or withdrawing.

inhibition, distress in response to pain, and shyness. The heritability of negative emotionality is particularly high, perhaps reflecting an unusually reactive amygdala. As a result, some children find it more difficult to learn to control their emotions than others do and may need extra help from you in reaching this goal. As we suggested in the previous chapter, the effects of your parenting may also be exaggerated in children like this (see also chapter 26).

As the cortex develops, around the middle of the second year of life, children begin to display the secondary emotions: pride, shame, guilt, jealousy, and embarrassment. These social emotions become more situation-specific as children come to appreciate the importance of their own behavior in causing particular

DID YOU KNOW? SELF-CONTROL PROMOTES EMPATHY

Babies often cry when they hear another baby crying, in the contagious form of empathy (see chapter 19). True empathy, the ability to appreciate other people's feelings, develops by age five. Children show great gains in self-control during that same period, and individual children who have better self-control also show more empathy and a more developed conscience. Similarly, children who are better at inhibiting an automatic behavioral response (for example, by saying "day" when shown a picture of the moon, instead of "night") tend to have a more sophisticated ability to imagine what other people are thinking and feeling, even when age, intelligence, and working memory are taken into account.

What's the connection? To develop empathy, children need to have a certain ability to entertain alternatives and hypothetical possibilities, which is part of self-control. The prefrontal and anterior cingulate regions of the cortex are involved in both self-control and empathy, so cortical development may limit children's ability to understand other people's feelings and their ability to control themselves. As part of the brain's attention-regulation system, the anterior cingulate is active when children concentrate on their own behavior or on other people's feelings. Such concentration may be a necessary first step in the development of both these important functions. The prefrontal cortex is important for behavioral inhibition, and it is active during theory-of-mind tasks, which require people to concentrate on what someone else could know.

outcomes. For example, at six, children feel guilty whenever something goes wrong, but by nine, they understand that guilt is only appropriate when their actions were responsible for what happened.

As we pointed out in chapter 10, the neocortex actively interprets external experiences, for instance, turning raw sensory information into a coherent sense of vision. As emotion-related cortical areas mature, they also gain the ability to shape how emotions are experienced. The anterior cingulate, insular, and orbitofrontal regions of the cortex combine numerous brain signals to construct the conscious experience of emotion. The anterior cingulate may be involved in seeking understanding or control over emotions. The orbitofrontal cortex is important for evaluating the social context of your choices, while the anterior insula represents your internal state, receiving signals from a wide range of brain areas about everything from thirst to cigarette craving to love.

The conscious perception of an emotional state is influenced by memories, by inferences about causality, by ideas about how to respond to the situation, and by the social context. For this reason, in addition to lacking control over their emotions, young children probably do not experience emotional states as richly as adults do until the cortex matures.

When babies first start to direct their own attention, by eight to ten months of age, they start to use distraction to manage their emotions, for instance, by turning to a new activity when a toy is taken away. Indeed, infants who can direct their attention to one particular thing for long durations express more positive emotions as babies and show better self-control later in life. As children's brains develop more self-control in general, their ability to regulate their emotions grows as well, as the same brain circuits are involved in all forms of self-control (see chapter 13). Children's brains apparently have to work harder than adults' when trying to inhibit an ongoing behavior. Older children gradually learn to use more sophisticated strategies, such as reinterpreting the meaning of an event to manage their emotional reactions ("That teacher doesn't hate me; she's just in a bad mood because Justin's been talking back to her all day"). As they get better at managing their emotions, children also improve their ability to hide emotions, allowing them to smile at Grandma for giving them a sweater—no matter what it looks like.

Parents who are more sensitive to their infant's needs and respond quickly to emotional cues tend to raise children who are better at regulating their own emotions.

Until their own regulatory capacity is fully developed, your children rely on you to moderate their emotions, by soothing or distracting them, and to help them learn how it's done. Parents who are more sensitive to their infant's needs and respond quickly to emotional cues tend to raise children who are better at regulating their own emotions. Maternal warmth and the strength of the bond between mother and child (see chapter 20) are also correlated with children's self-control ability. In other words, a good relationship with Mom may be a long-term source of willpower. Finally, parents who explicitly coach their children on the experience of emotions, by labeling and validating feelings and suggesting constructive ways to deal with them, tend to have children who are better at regulating their emotions later in life. Since children with poor self-control ability are prone to aggression and are at risk of developing conduct disorders and ADHD (see chapter 28), helping them self-regulate can do them a great service.

The prefrontal cortex, the newest part of our brains in evolutionary terms, is very slow to mature, developing last in the back-to-front progression of the neocortex (see chapter 9), but its capacities are worth waiting for. During the first two years, its neurons increase their complexity and add many synapses. After that, this part of the cortex enters the stage of experience-dependent synapse elimination (see chapter 5). Its connections do not become completely adultlike until late adolescence, and long-distance connections through white matter develop even more slowly. So your child's ability to regulate his emotions and to recognize and respond appropriately to the emotions of others continues to improve throughout childhood. That promise should give parents something to look forward to on those days when life with a toddler is just plain exhausting.

Chapter 19

EMPATHY AND THEORY OF MIND

AGES: ONE YEAR TO FIVE YEARS

The first time your child tries to pull the wool over your eyes, take a minute to appreciate the achievement. Although we want children to tell the truth, gaining the ability to lie convincingly is a notable step in mental development. Even to attempt it means the child thinks she can manipulate an adult's knowledge. It also means that she knows that others have thoughts, some of which contradict reality. Understanding that others can have false beliefs is part of normal development—and appears to be unique to people.

People are intensely social animals. We form alliances, jockey for position, comfort one another, and play games. When we're not doing that, we talk about each other, speculating on one another's motives. All these activities require a capacity to think about other people's state of mind. Here we will describe how these brain systems come online—as well as give you some tips on watching these qualities grow in your son or daughter.

This ability, called **theory of mind**, involves several components that appear in stages. As early as three months, infants divide the world into objects and agents (see chapter 1). Their ability to identify agents, which act with purpose and goals, provides a foundation for understanding other minds.

Theory of mind can be sorted into two categories, which rely on different but overlapping brain systems. Over the first two years of life, children become able to react to other people's perceptions and desires. In the second phase, also lasting about two years, children gradually develop a system for thinking explicitly—and talking—about the beliefs of others. By around age four, children possess a full-blown theory of mind.

The understanding of feelings can occur at different levels. The brain has a system for rapid processing of emotional stimuli, which reacts in a contagious manner to the feelings of others. This process involves the sensory neocortex, the thalamus, and the amygdala. With enough time, such processing can also involve the neocortex, including temporal and frontal areas, in a cognitive form of empathy.

Contrary to the wisdom of some folk psychology, many animals have the contagious form of empathy. Rhesus monkeys will refrain from pulling a chain that delivers an electrical shock to another monkey, even if they know that pulling the chain will bring them a large food reward. Rats can help each other as well. When confronted with a squealing, wriggling rat suspended in a sling, another rat will press a lever repeatedly to bring the suspended rat to the floor, staying close to it the whole time. So as originally suggested by Charles Darwin, empathy in its most basic form, the desire to assist members of one's own species, is widespread among mammals.

Apes, which have a larger neocortex than rats or monkeys, are capable of more cognitive acts of empathy. For example, chimpanzees will comfort sick and injured birds, gently straightening their wings to aid their attempts to fly. (Showing a lack of foresight, they sometimes then toss the birds from high places.) They are also reported to help unfamiliar people in distress.

As the neocortex develops, children make the transition from rat- and monkeylike empathy to apelike empathy. As many parents and caregivers know, the presence of one crying baby in a nursery often starts the others crying. As children grow older, empathetic responses become more complex. Children imitate the distress behaviors of other children, as if trying them on to see how they feel. They soon shift away from feeling personal distress and start showing helping behaviors. In the second year of life, toddlers comfort younger siblings in distress by patting, hugging, or kissing them. Similarly, they may bring a security blanket to an adult in pain.

How does the baby's brain generate contagious empathy? Many scientists believe that recognizing emotions in others involves experiencing the emotion yourself. Indeed, scans of brain activity in adults show that the insula is activated when they experience feelings of disgust—and when they look at disgusted faces (see *Did you know? Imitation in the brain*, p. 168). Similarly, the amygdala is activated when we look at frightened faces and also when we feel fear ourselves.

By internalizing the emotions of others, children develop a sense that others have desires. This appreciation is clearly visible at twelve months; if a baby sees someone look at an object with positive emotion, she expects him to reach for it—and looks for a longer time if he does not. (This measure of violated expectations, spontaneous looking when something unexpected happens, is one of the ways to measure infant abilities. For more on this topic, see chapter 1.) Between fourteen and eighteen months, babies acquire the capacity to understand that when a grown-up has indicated a preference for a food, the baby should give it to him—even if the baby doesn't like that food himself. You can see this when your

DID YOU KNOW? **OLDER SIBLINGS SPEED A CHILD'S THEORY-OF-MIND DEVELOPMENT**

The ability to reason about the beliefs of others depends on brain maturation, but it is also influenced by experience. The more often a parent talks to a child about motivations and mental states, the sooner the child begins to demonstrate an explicit ability to speak in terms of the beliefs of others. An even stronger influence is growing up with older siblings. Having an older (but not younger) sibling speeds the onset of theory-of-mind capacity in three- to five-year-old children. The size of the difference is equivalent to an average of four to six months of advance per older sibling, for up to three siblings. By age six, nearly all children have acquired the same level of understanding, but there may be lasting social advantages to developing this capability earlier (see chapter 20). So in the preschool years, living with older siblings can help a child's mind mature faster.

toddler hands you a piece of asparagus (assuming you like asparagus). Children also begin to have some sense of what others feel about them, as shown in the development of self-conscious emotions such as embarrassment during the second year (see chapter 18).

In parallel with these social emotions, babies between the ages of thirteen and twenty-five months also begin to develop a framework for understanding the perceptions and desires of other people. In one study, babies of twenty-five months saw a bear puppet stowing a toy inside one of two boxes for an observer to retrieve. Then, after hiding the toy again, the bear switched the toys while the observer was looking the other way. When the observer turned back, the children looked at the original box, as if expecting the observer to look in the wrong location. In other tests, young children also looked longer when people acted as if they knew something that they were not supposed to know, such as the location of a tasty snack that had been covertly moved. Putting on pretend scenes like these can be amusing to you and your child—and educational for you.

This level of sophistication is unique to people, but chimpanzees come close. They understand the goals and intentions of others, for instance, reaching for food that another chimp cannot see instead of food that is visible to both. But researchers have not yet found a way to get chimps to act on the wrong belief of another chimp or human.

Preschoolers have one more hurdle in their development of theory of mind: they have to learn to verbalize their awareness of another person's false belief. This ability arises well after children begin to talk. In one classic test of theory of mind, a child is told a story of two girls named Sally and Anne. Sally has a basket with a lid, and Anne has a box. While they are together, Sally puts a marble in her basket. Sally leaves, and while she is away Anne moves the marble to the box. When Sally comes back, where will she look for her marble? Most preschoolers will indicate Anne's box, where the marble really is. Only around the age of four will children start indicating the basket, where Sally is likely to wrongly *think* her marble still is.

The ability to think about the thoughts and beliefs of others probably grows from earlier, more basic capacities. Activity measurements from the brains of adults provide an indication of where those earlier capacities might be localized. Parts of the neocortex near the posterior superior temporal **sulcus** are activated when people detect an agent, and the temporal poles of the neocortex are active when social knowledge is generated. The inferior parietal lobule is activated by visual motion or the direction of another person's gaze, which conveys information about that person's intention.

Gaining the ability to lie convincingly is a notable step in mental development

A key component of theory of mind in the brain is the medial prefrontal cortex. This region is active when people are asked to distinguish someone's mental representation from a real-world physical state. The medial prefrontal cortex is also activated during humor, embarrassment, and other moral emotions. This region develops relatively late, as does long-distance communication between frontal and temporal regions, perhaps accounting for why these capacities only come together starting around age four.

Around age two, children start to play simple pretend games. A year later, they show a more elaborate understanding, doing things like putting on a raincoat and boots in sunny weather, then pretending to splash in puddles. The developmental psychologist Alison Gopnik calls these *silly mental states*—a third state

DID YOU KNOW? **IMITATION IN THE BRAIN**

Sam once stuck out his tongue, with the sides curled up, at his baby daughter, and she was able to mimic him perfectly on her first try. This was quite a surprise. Such a complex act of imitation meant that she was able to generate a strange new facial expression simply based on seeing the movement—a complex mapping of a visual stimulus to a coordinated set of corresponding tongue muscle movements. Of course, you might be less of a geek than Sam, and just like the sight of your baby sticking her tongue out.

How can this imitation happen? Across many brain regions, direct experience activates some of the same neurons as vicarious experience. In electrical recordings from the **premotor cortex**, a frontal brain region, there are neurons that are activated when a monkey makes a specific movement, such as grasping a piece of fruit to bring it to the mouth. Researchers found that particular neurons, which they called *mirror neurons*, fired both when the animal performed such a movement and when the animal saw the same movement performed by someone else. Mirror neurons for specific actions are also found in the brains of people undergoing exploratory neurosurgery.

Mirror neurons have received a fair bit of hype in the popular press as mystical causes of a variety of other capacities, such as empathetic abilities. Although much of this discussion is overblown, there is a more general principle regarding how other people's emotions and actions are represented in many areas across the brain.

For example, emotion-related brain areas show mirrorlike properties. Both strong negative emotions and the sight of a face expressing the same emotion trigger activity in the insula, a cortical region that communicates with other emotion-processing regions such as the amygdala. The insula also receives information from the premotor cortex, suggesting that mirror neurons could convey the emotional content of body language. Information goes in both directions between these brain regions, so they could even teach one another about emotions and their physical expressions.

Mirror neurons are just one example of how single neurons can represent surprisingly abstract concepts. The inferior temporal cortex contains neurons that fire selectively in response to faces, body parts, other objects,

the memory of recent objects, and even a person's identity. Researchers recorded a single human neuron that responds both to an image of the actress Halle Berry and to the letters H - A - L - L - E - B - E - R - R - Y. Recognizing a celebrity may not have the same social importance as learning empathy for other people, but the abilities of the neurons are just as amazing.

---

of mind, separate from knowledge of the world and absolute ignorance. Silliness is pretense chosen on purpose, just for fun. Your children exercise this sense when they watch a puppet show and shout at the hero or heroine to watch out, or when you pretend not to know something. The ability to reach this state—and perhaps see it in playmates—may be a stepping-stone to a full-blown theory of mind.

Children at this age can remember what they were pretending before and have conversations with themselves in the present. Strangely, though, they cannot remember their previous mental states. For instance, if you give a hungry child his lunch, by the time he is done he is unable to recall that before eating, he was hungry. Likewise, if you play a game with a pencil box in which you surprise the child with crayons instead of pencils, once he has seen the crayons he cannot recall once thinking that there were pencils in the box. Adults also can have difficulty remembering their past mental states, such as whether they liked a roller coaster while they were on it. Perhaps you know one or two people who are like this.

Thinking about one's past or future self and thinking about other people are closely related. Both capacities involve self-projection—the ability to think about mental states that are not currently your own and to shift perspectives in general. These capacities require the most frontal parts of the brain, which are the latest to develop.

Frontal brain development has additional profound consequences. Children begin to see events from multiple points of view. Also, we learn to understand ourselves at least in part by observing others and their reactions to us. Finally, the ability to attribute feelings and intentions to others is an essential component of mature religious beliefs, which depend on faith in unseen motives. Although fibbing is socially and often practically undesirable, it also represents the opening of new vistas for your child's mind.

Chapter 20

PLAYING NICELY WITH OTHERS

AGES: BIRTH TO EARLY TWENTIES

Those moments when you feel in sync with your baby, as you trade gestures, facial expressions, words, or just silly noises back and forth, are crucial for early brain development. To you, it's a fun game of peekaboo. To your baby, it's an education in self-control, as well as his first experience of relationships.

A sensitive adult can regulate the baby's arousal level, which young infants can't do for themselves, by responding to cues (such as turning toward the partner) that indicate when the baby wants more interaction, and to other cues that indicate the interaction has reached its best intensity (smiling) or that the baby is overstimulated (looking away). Even sensitive parents frequently misinterpret their baby's cues, so that the baby gets a mixed experience of synchrony and missed connections. This seems to be optimal for development.

Babies reach out for this type of interaction thousands of times a day. Young babies become quite upset if their partner briefly stops responding by freezing in place. When this happens, the baby's heart rate and stress hormone levels both increase. A lack of response is more upsetting to infants than physical separation from the mother or having her turn away to talk to someone else. Even after the partner starts responding again, babies are fussier and less responsive to interaction for a while. After six months, they slowly become better at coping with unreliable responses from their partners and more able to regulate their own arousal.

Problems affecting either parent or infant can interfere with the development of synchronous behavior. Depressed mothers respond more slowly to their babies, are less sensitive to their cues, and show less affection than nondepressed mothers.

Their babies are less responsive to faces and voices and show less distress when their partner stops responding to them than the babies of nondepressed mothers. Premature babies' brains are not mature enough to support normal synchronous interactions, so they are more irritable, signal less clearly, and have more trouble tolerating mismatched interactions. These deficits may be important because better synchrony in infancy predicts more secure **attachment** at one year of age, better self-control at two through six, and more empathy at thirteen.

Attachment is just what it sounds like: a strong and persistent desire to be close to a familiar caregiver, especially when the child is stressed or upset. Infants do not attach to their parents until about six months of age. The delay may seem curious, since parents begin to form attachments much earlier, within a few weeks after birth. But it makes sense from an evolutionary perspective: babies are safer if they are around a protective adult. Young babies pretty much stay where you put them, so at first it's more important for the parent to attach. Attachment on the part of the baby serves a protective function only after he can move on his own. Before this age, it appears that the frontal cortex is not yet mature enough to allow the baby to form attachments.

Most babies form attachments to more than one person, though the strongest attachment is usually to the mother. Synchronous interactions contribute to the development of attachments. All babies with normal brains and consistent caregivers form attachments, but the quality of that attachment varies quite a bit. Babies who are securely attached (over half of the population) use their mother as a base for exploration, periodically keep in touch with her as they play, are mildly distressed when she leaves, and are comforted when she returns. Babies with the insecure-ambivalent pattern (under a quarter of the population) stay close to their mother when she is present, are extremely distressed when she leaves, and are alternately clingy and angry when she returns. Babies with the insecure-avoidant pattern (about a fifth of the population) show little apparent distress when the mother leaves and refuse to engage with her when she returns. A rare fourth category, disorganized attachment, is characterized by inconsistent behavior and occurs mainly in babies who have been abused or neglected. Babies may show different attachment styles with different caregivers, but by age five children typically use a single dominant style, which is moderately stable for the rest of their childhood.

By nine months, as babies develop the ability to direct their attention, they become better at initiating synchronous interactions, and they start to experience stranger anxiety, becoming uncomfortable around unfamiliar people. Babies also begin to draw their partner's attention to objects in several ways: by smiling at the object and then smiling at the person, or by pointing at the object, or by looking alternately at the object and the person. This behavior is more common among infants who respond, when their mother stops interacting, by smiling and making other attempts to engage her socially. Initiating joint attention to an object is one of the earliest indications of social skills, and babies who do a lot of it at nine or ten months are more likely to be rated as socially competent at thirty months.

The outcome of these early social interactions cannot be pinned on either inborn tendencies or parenting style alone. Because social interactions are reciprocal, your baby influences your behavior as much as you influence hers, making it difficult for researchers to sort out the causes of particular interaction patterns. Babies who respond positively to care are likely to receive more attention than babies who are less happy or are difficult to comfort, probably because their parents find taking care of happy babies more rewarding. For this reason, many parent-infant interactions are self-reinforcing, either positively or negatively. For instance, infants who cry more at six months end up with mothers who respond less at twelve months, but mothers who respond more to their infants at six months end up with infants who cry less at twelve months.

We know that synchronous interactions are important to attachment formation because interventions to improve maternal sensitivity increase the probability of secure attachment. Several other factors also contribute importantly to attachment formation, including physical contact, socioeconomic status (see chapter 30), and temperament. For example, in a study of economically disadvantaged mothers, babies were more likely to form secure attachments if the mothers used a harness to strap the baby to their bodies than if they used a car seat to hold the infant.

Because social interactions are reciprocal, your baby influences your behavior as much as you influence hers.

Notably, for most children, the evidence shows that going to day care does not interfere with secure attachment. A longitudinal study in Sweden found only positive effects of day care, though a similar U.S. study found that children with insensitive mothers were more likely to form insecure attachments if they also went to day care before one year of age.

Several long-term studies found that secure attachment in early childhood predicts social competence and better social outcomes in later childhood, particularly in children with a shy or inhibited temperament. For example, attachment security predicts self-control much more strongly in children with a short version of the 5-HTT serotonin receptor (see chapter 26) than in children with a long version. The effects of attachment security on later life are small to moderate, so other factors contribute importantly to socialization as well.

Social competence is built on a foundation of basic skills, including self-control (see chapter 13), emotional maturity (see chapter 18), and theory of mind (see chapter 19). All four abilities develop along a similar time course and tend to track together within individuals as well, though with some variability. The

DID YOU KNOW? STEREOTYPING AND SOCIALIZATION

The way that the brain produces a sense of group membership has a dark side. Social groups are defined not only by their members but also by the people who are outside the group. Especially in situations where groups are competitive, your brain easily attributes negative characteristics to the opposition, the way that Duke basketball fans have a tendency to denigrate the University of North Carolina. This hostility isn't necessarily restricted to the players, or the basketball program, and may generalize to denunciation of the entire university.

As we discussed in chapter 1, your child's brain is naturally inclined to put things and people into categories. Young children tend to use the most visible characteristic to sort out categories. For this reason, young children are sensitive to superficial group differences like race, whether or not they are explicitly pointed out. This sensitivity starts to emerge during the preschool years, when children begin to socialize in groups. A difference as meaningless as giving children shirts of different colors can create a feeling of membership that causes them to like children within their own group better than children of the other group. Unless parents actively teach children not to attribute characteristics to people by race, they are likely to socialize preferentially with others of their own race. This pattern persists throughout the school years even among children who go to mixed-race schools and day-care facilities.

similarities probably occur because these abilities are limited by maturation of some of the same brain regions, particularly the anterior cingulate and prefrontal cortex, the latest-developing part of the brain. Social cognition also involves many other regions of the cortex, including the posterior superior temporal sulcus, the temporoparietal junction (the area where the temporal and parietal lobes meet), and the anterior insula, as well as the amygdala.

As children grow older, they socialize outside the family more and more. At all ages, children are most likely to choose friends whose characteristics are similar to their own. The earliest peer relationships in toddlers are one-on-one interactions characterized by turn-taking and mutual imitation, an early form of cooperation, and by frequent conflict, usually over toys. Group interactions are not well developed at this age. Older preschoolers increasingly participate in imaginative play and games with rules, both of which require the prefrontal cortex. Helping and sharing become more common during the preschool years, and aggression declines after age three. At this age, conflicts are more likely to involve ideas and opinions than struggles over things, and language becomes progressively more important to social relationships. Social competence is already associated with social success at this age, probably through a feedback loop in which better socialized children make more friends, which allows them to improve their social skills still more.

Successful socialization involves both formation of individual friendships and acceptance into peer groups. Friendships are the major source of affection, while group interactions are the main source of power and status. These two forms of socialization are both associated with good psychological health across cultures, though different societies may emphasize one aspect more than the other.

Once children enter school, their peer interactions become more frequent and less closely supervised by adults. In this age range, verbal aggression (threats, gossip, and insults) largely replaces physical aggression, and positive interactions also increase. Hostility begins to be expressed as persistent dislike of a particular person, rather than being restricted to a situation. The frequency of rough-and-tumble play peaks during the early school years. Competitive interactions via formal or informal games become more common. Children's concepts of friendship become more sophisticated, moving from shared activities in preschool to shared values, self-disclosure, and loyalty by early adolescence. Social interaction between boys and girls drops off sharply around age seven (see chapter 8) and resumes again in early adolescence. Almost all children in this age range are members of a group of three to nine children who rarely play with anyone outside the group.

As children near adolescence, their group memberships usually become more fluid, as they interact with a larger number of other people in a variety of contexts. At this age, attempting to figure out what other people think and feel requires more prefrontal cortex activity than the same task in adulthood, as these parts of the brain continue to develop until the late teens (see chapter 9). Romantic relationships start to appear by age twelve, and their duration and frequency increase through the teen years. The quality of earlier friendships, particularly those involving negative interactions, moderately predicts the quality of later romantic relationships.

Both the major components of social competence—sociability and behavioral appropriateness—are strongly influenced by the culture in which your child is raised. Within a given culture, certain children are more inclined to shyness than others, probably due to genetic biases, but identical twins raised in different cultures behave differently. Societies that encourage social interaction (such as the U.S. and Italy) produce children who tend to be less shy or inhibited than children whose societies value modesty and cautiousness (such as rural China and India). These differences can already be observed in toddlers. In the former countries, shyness (especially in boys) is met with disapproval or punishment from parents and rejection from peers, while in the latter countries parents and peers react to shyness with acceptance and approval. These differences mean that extroversion gives people an advantage in some societies but not in others. For example, in the U.S., shyness predicts lower educational and professional achievement, but in Sweden shyness has no such effect.

The value placed on behavioral appropriateness also varies across cultures. In traditional, agrarian societies made up of extended families, group harmony is considered very important, and children are mainly cooperative and compliant. Mothers in these cultures spend a lot of time in physical contact with their young children, expecting and receiving a high level of obedience. Older children, beginning as early as age five, are assigned household tasks, and their responsibilities increase substantially as they get older, to as much as six to seven hours per day at age ten to twelve.

In urban and industrial societies, individual accomplishment and competition are more highly valued, and children are more likely to be defiant or aggressive. In general, displays of anger toward other people are more common in societies where parents respond to them by trying to coax the child into feeling better and less common where parents express disapproval of such behavior. Children also respond to their peers' attitudes about aggression, which may increase children's status in individualistic cultures but decrease it in group cultures.

Of course, not all peer relationships are positive. About a third of school-age children report having a relationship characterized by mutual dislike, which can have significant developmental effects. Children with such negative relationships are more likely to have problems, ranging from depression and withdrawal to aggression. They are also less likely to do well in school and more likely to experience a variety of difficulties with their peers, including being bullied, being unpopular, and having trouble forming friendships. Girls and boys are equally

PRACTICAL TIP: **PROMOTING CONSCIENCE**

The first stirrings of moral behavior in children come from the brain's emotional system, which remains an important contributor to our moral sense in adulthood. Some precursors of the moral sense are probably built into the brain—even young babies like agents who help others better than those who hinder (see chapter 1)—but parenting has clear effects on its development.

The earliest precursor of conscience is your child's desire to please you, which tends to be stable across situations, whether you're teaching her to count or asking her not to write on the walls. Before age two, children begin to show individual differences in their likelihood of feeling guilty when they've done something wrong, which is linked to their ability to follow rules when no one is watching. Receptiveness to parental guidance predicts individual differences in conscience at later ages, including older children's ability to reason about moral situations.

You might think that children would be more likely to obey strict parents, but this approach is actually more likely to produce a rebel. Parents who repeatedly assert their power interfere with the development of guilt and later conscience, producing children who blame external factors for their faults. The parents who receive the most obedience are the warm ones, whose children comply willingly with their parents' wishes from a desire to make them happy (see chapter 29). Mutually positive parent-child interactions are a strong predictor of later moral behavior, particularly in securely attached children.

As you might expect, your child's characteristics also influence the development of conscience. Children with strong self-control show more mature moral abilities than impulsive children of the same age. Children who are temperamentally fearful are prone to guilt, which leads to more compliant behavior, so for them warm and sensitive parenting is the most effective path to conscience, as the development of excessive guilt can lead to later anxiety disorders. For children who are less fearful, attachment security, which leads to an interest in pleasing the parent, is the best predictor of later conscience.

likely to have negative relationships, and these relationships are equally likely to be between children of the same sex or different sexes.

Social withdrawal in childhood may also lead to trouble. On average, behaviorally inhibited children show higher than normal activity in the right frontal cortex, the area associated with emotions that lead to withdrawal (see *Myth: The right hemisphere is the emotional side*, p. 158). Children who are reluctant to interact with other children miss opportunities to practice their social skills, so the condition is often self-perpetuating. In the U.S., for example, social withdrawal at age seven is a risk factor for depression, loneliness, and negative self-image at fourteen. Shyness is equally prevalent in boys and girls, but the costs are higher for shy boys, who experience more stress and more peer rejection than shy girls, presumably because of different cultural expectations for male and female behavior. About a quarter of socially withdrawn children are targeted for bullying. (For comparison, about half of overly aggressive children are targeted.)

What can you do to help an inhibited child to avoid these problems? As we said earlier, warm and sensitive parenting with lots of physical contact promotes secure attachment, which predicts improved social competence for inhibited children. It's not helpful to micromanage your child's behavior; that can interfere with the development of social skills. But gentle encouragement to join groups of other children can be useful. Children who rarely express negative emotions like sadness and anxiety to their peers tend to be better accepted and less likely to be bullied, so improving emotional regulation might be one useful avenue to explore. Poor emotional regulation predicts poor social adjustment across the lifespan. Participation in organized sports or other extracurricular activities may be helpful, as talents valued by other children also promote peer acceptance. Early intervention is preferable because it is easier for children to catch up when their social skills haven't yet gotten too far behind those of other children of the same age.

In all cases, building a connection with your children while respecting their individual temperaments provides a solid foundation for socialization and many other forms of learning. As with most aspects of child development, some factors that contribute to your child's socialization are beyond your control, but there are ways that you can help. Anyway, building a warm relationship with your child should be a worthwhile goal in itself.

PART SIX

YOUR CHILD'S BRAIN AT SCHOOL

Chapter 21

STARTING TO WRITE THE LIFE STORY

AGES: TWO YEARS TO EIGHTEEN YEARS

Considering that young brains are so good at learning, it's odd that children can't remember what happens to them in the first few years of life. You know that memorable episode involving chocolate pudding, your fancy clothes, and your two-year-old? He won't recall it when he's ten. What's going on here?

The brain has many different types of learning, only one of which is dedicated to facts and events that we can consciously recall. Early experience makes a strong impression on brain function in many ways, as we discussed in the first few chapters of this book. In contrast, memory for events develops comparatively late, so the life story does not start to be reliably recorded until age three or four.

All kinds of learning are driven by cellular mechanisms that alter neurons and the synapses that connect them. Children and adults undergo similar cellular changes during everyday experience, but they function somewhat differently in children. The properties of these learning mechanisms may explain why we never forget how to ride a bike and why taking breaks from study can aid learning.

Different forms of memory call upon different brain regions. In adults, the principal distinction in memory type is the difference between *declarative memory* and *nondeclarative memory*. Declarative memory is the recall of a fact or an episode from the past. Formation of declarative memories requires the hippocampus, a horn-shaped structure found on both sides of the brain (see figure, p. 126) that is also necessary for spatial navigation. Another key region for declarative memory

is the neocortex, especially its medial temporal regions. Declarative memory continues to improve throughout childhood.

It is possible to see glimmerings of hippocampus-dependent memory before age two. For example, your preschooler might be able to describe an episode that happened to her before she could speak. Even so, memory is difficult to probe in the very young because children cannot declare much until they have a good command of language. For this reason, compared with nonverbal measures, verbal measures of memory underestimate memory competence in children up to age six.

A better way to examine memory during this period is to observe spontaneous actions (see chapter 1) or to give children a task that does not require speaking. For example, in the laboratory, children of eighteen to twenty-four months can learn to navigate with the help of distinctively shaped objects as landmarks to remind them where to go. In real life, your child may complain if you take an unusual route to school or day care. These behaviors are evidence of memory for the locations in question.

The other major category of memory is nondeclarative, encompassing a wide range of nonverbal memories: learned associations, skills, and habits. One type of nondeclarative memory that is readily studied in infants is the formation of associations: for example, anticipating that Mom is about to leave the house when she puts on her coat. Associative learning is present from a very early age, though infants also forget associations relatively quickly (see *Did you know? Babies forget faster*, p. 186). Associative learning can involve the amygdala for emotional responses or the cerebellum for other forms of sensory learning.

Infant remembrance can be boosted with timely reminders.

Another type of nondeclarative memory is *procedural learning*, the acquisition of habits and skills, such as learning to tie your shoes. Shortly after the first birthday, children can perform procedures in a sequence, such as making a rattle by putting a ball inside a container, putting the lid on, and shaking it. This procedural learning requires the striatum, a structure in the basal ganglia that is necessary for movement and initiating actions. Procedural skills are learned more robustly than declarative memories requiring the hippocampus. This is why you never forget how to ride a bike.

To learn the new, an infant must get used to the old. As described in chapter 1, infants are good at detecting new information in their environments. A necessary part of this capacity is the ability to identify objects as familiar. Infants less than one year old will look with interest at a new object revealed when a cloth is pulled away. With successive repetitions, they slowly lose interest and eventually stop orienting altogether, a process called *habituation*. After a waiting period of minutes, the baby recovers from habituation and looks with renewed interest. Some of this memory is longer lasting; if you start over again, it doesn't take as long for the infant to habituate as it did the first time. Eventually the infant ignores the object entirely.

Habituation does not initially require long-term changes in the brain. It is present in nearly all animals, including sea slugs, fruit flies, and even single-celled organisms. Habituation depends on short-term changes, such as the accumulation or depletion of intracellular chemical signals, which then inhibit neurons from firing or prevent synaptic terminals from releasing neurotransmitter.

Similar chemical signals are also triggered by new experiences. These signals can cause long-term changes in the structure and composition of neurons and synapses—the nuts and bolts of learning. Most of these changes are unlikely to include the generation of new neurons. Nearly all neurons that the brain will ever contain are already present soon after birth (see chapter 2). After that, new neurons are produced only in the olfactory bulb, part of the hippocampus, and at a trickle in the neocortex. Also, because neurons have formed most of their axons and dendrites within the first few years of life, the possible locations where a neuron can form new synapses are also somewhat constrained.

Given these commitments, a major site for learning to occur in older children and adults is within existing neurons and at the synapses between them. Existing synapses can grow stronger or weaker, as neurotransmitter receptors are added or removed, or as the neurotransmitter chemical becomes more or less likely to be released at the connection when the presynaptic neuron fires. In response to external events or internal processes, information in the brain flows along paths of neurons that fire with characteristic patterns and sequences.

The process by which individual synapses become stronger is called *long-term potentiation.* It requires particular patterns of neuronal firing as well as signals such as dopamine, acetylcholine, and other long-distance neurotransmitters. When these conditions are met, a group of neurons firing together in sequence can trigger biochemical processes that strengthen the connections between all the neurons. As we mentioned in chapter 5, cells that fire together, wire together.

Just as important in learning is the weakening of synapses. Synapse strength is reduced in a process called *long-term depression.* This process occurs when two connected neurons fire independently of one another, which can happen when the postsynaptic neuron is being driven by other neurons. Or, in a slogan often used to teach neuroscience students, "Out of sync, lose your link." In addition, as we saw in chapters 5 and 9, childhood is a period of ongoing synapse elimination, in

PRACTICAL TIP: THE BEST STUDY HABITS

Decades of research have identified study techniques that can vastly improve learning, but most teachers don't practice them, and few parents know about them. Fortunately, these strategies aren't complicated, and your child can use them at home.

Students often wait until the last minute, then make up for lost time in a marathon study session. This approach flies in the face of one of the most reliable results in research on learning: the power of spaced study. The brain retains many kinds of information longer if there is time to process the learned information between training sessions. Two study sessions with time between them can result in twice as much learning as a single study session of the same total length. Spaced training works with students of all ages and ability levels, across a variety of topics and teaching procedures. In general, the longer the gap between study sessions (up to a year in some cases), the longer people will remember the material.

One possible reason why this approach works is that breaks allow time for newly acquired information to be consolidated (see p. 187). Memories are not written just once but reinforced either during recall or even offline, for instance, during sleep. Both declarative and procedural memory appear to be consolidated during sleep, so it's important to make sure your child gets enough rest.

Because memories are reconsolidated when they are recalled, tests actually improve learning (and slow down subsequent forgetting) by compelling the student to actively recall the course material. Passive reading is much less effective for learning. Multiple-choice tests do not improve learning, while short-answer questions do. You can take advantage of this fact by quizzing your child at home during study time to improve her performance at school and by teaching your child to test herself as a study strategy.

A third way to improve your child's learning is to mix it up. Children who see ten similar examples in a row learn considerably less than children who see ten different examples. This strategy works across domains, affecting the way we learn sports, art history, math, or any other subject. Varying the timing and location of study sessions also improves recall, probably because

learning is contextual, so learning in multiple contexts gives your child's brain a deeper connection to the material.

At first, your child may find these approaches discouraging because they often result in more errors during studying—but they will produce better test performance with no more effort than traditional study techniques. Such good results should change his tune quickly.

---

which many synapses disappear entirely as a part of normal development—the refinement of your child's brain circuits. Synapse elimination is driven not only by activity but also by the availability of trophic factors, proteins that are necessary for the growth and maintenance of dendrites and axons.

The formation and breakage of synapses is another major site of information storage. In most of the brain, neurons form synapses with only a small fraction of their neighbors, so the growth and elimination of connections can establish entirely new pathways for information. This process is happening on a very large scale during childhood, when synapses are produced in large numbers, and then eliminated according to experience (see chapters 5 and 9). In addition, neurons can change how they respond to synaptic input, for instance, by changing their electrical and chemical response properties. All of these processes require new proteins and cellular structures to be made and broken down.

Tests improve learning by compelling the student to actively recall the course material.

You may know that Paris is the capital of France, but you are far less likely to recall where you were when you learned this fact—unless you learned it recently. In contrast, your child might be able to recall where he heard it. The conversion of short-term to long-term memory seems to involve physical transfer from one brain region to another. Initial storage of a fact requires the hippocampus and other brain structures nearby, in medial temporal parts of the neocortex. The hippocampus sends connections to the neocortex, and with time, factual information is reprocessed to join our storehouse of general knowledge. The relatively late development of synaptic connections in

DID YOU KNOW? **BABIES FORGET FASTER**

For a long time, *infantile amnesia*—the near-absence of memories from before one's third birthday—was interpreted to mean that infants only have a primitive capacity to form memories. But young children can recall things that happened earlier, suggesting that information is stored but gets lost on the way to adult life. One possibility is that the brain is incapable of transferring memories into long-term storage in the neocortex. Recent evidence suggests that an alternative cause is instability of the initial memory.

As we described in chapter 1, infants can learn to form associations, as evidenced by their ability to learn to kick when their foot is attached by a ribbon to a mobile. When they get older, they outgrow this game, so that researchers have to come up with something more complex, such as pressing a lever to cause a miniature train to go around a track.

However, associative learning does not last long in infancy. At two months of age, babies remember for only a day. By three months, the duration increases to a week. After that, remembering grows steadily, until at eighteen months, children can remember simple associations for three months. With time, babies also develop another form of memory, the ability to remember complex actions that they've observed and imitate them later. Six-month-olds can reproduce a facial or body movement after a day. At eighteen months, they can make a more complex movement, like dressing a teddy bear, after a delay of four weeks.

Infant remembrance can be boosted with timely reminders. A six-month-old baby's performance on the miniature train task lasts only two weeks after a single day of training, but a single additional session with the train doubles the retention time. Four reminders spread over six months lead babies to remember the task for a full year. Reminders are effective even if the initial association appears to be forgotten. As memories fade, reminders that are similar but a little bit off can distort the original memory, as also happens in adulthood. These findings raise the interesting possibility that appropriate reminders of an event that happened in early infancy could enable recollection of the event much later—perhaps even as an adult.

the hippocampus may be the reason that children have poor declarative memory in their early years. For other types of memory, the stored information may be transferred between brain regions in an analogous manner.

Synapses are modified not only when we encounter new information but also later, as memories are reprocessed. Perhaps surprisingly, memories appear to be rewritten frequently. Unlike a computer's memory, a biological memory is reinforced by recall. It is as if the ink on a printed page got darker when the page was read. This process is known as *reconsolidation*, in which a stable (consolidated) memory is restrengthened. Changes can even happen offline when we are not actively thinking about the information, as memories are also strengthened during sleep (see chapter 7).

These changes in synapses and neurons participate not only in the learning of facts in school but in all changes in the developing brain. As it matures, your child's brain undergoes transitions that go well beyond what we think of as learning. Socialization, the development of motor skills, and long-term changes in behavior and attention all rely on the fact that the brain is plastic, as inborn developmental programs and experience work together to shape your child's brain.

Chapter 22

LEARNING TO SOLVE PROBLEMS

AGES: TWO YEARS TO EIGHTEEN YEARS

If your child believes that intelligence is a fixed characteristic, that belief will make her act less smart. Children who think a test measures their innate competence do not try as hard or perform as well as those who think that effort is the major determinant of success or failure. Because children who believe intelligence can't be improved tend to see failure as a sign of low ability, they are likely to give up in shame when faced with a challenging task. In contrast, children who believe that hard work can improve their cognitive abilities often welcome difficult tasks and bounce back from failure, feeling that they have learned from the experience. For this reason, emphasizing the importance of intelligence to children may paradoxically reduce their chances of success.

Accordingly, interventions to change students' views of intelligence can improve academic performance. In one longitudinal study, math test scores were static over two years in students who entered seventh grade believing that intelligence is fixed, while scores improved over time in their peers who believed that intelligence is influenced by experience.

The researchers then went to a different school and offered seventh graders an eight-week class (half an hour per week) on brain function and study skills. One group's lessons included the idea that intelligence can be modified through practice, which leads to the formation of new connections in the brain. The other group got a lesson on memory. Later, the first group scored significantly higher on math tests than the second group, though the two groups had performed similarly before starting the class.

Parents can encourage their children to handle failure constructively by prais-

ing them for what they do rather than for what they are. Though telling a child that he is smart or artistic or athletic may seem like a good way to make him feel good about himself (see *Myth: Praise builds self-esteem*, p. 254), it also teaches him to view those characteristics as fixed traits. On the other hand, praising your child for effort or improvement, or for choosing a particular way to respond to a problem, communicates that his behavior and choices are what matter to you. Since he can control his behavior but not his traits, that message is more empowering. (It's also great to communicate that you're on his team no matter what, but there are better ways to do that, such as saying "I love you.")

PRACTICAL TIP: SOCIAL REJECTION REDUCES IQ

Because intelligence is so closely connected to working memory, a lot of events that distract you can make you temporarily less smart. In particular, unpleasant or awkward interactions with other people can greatly influence performance on tests of cognitive function.

In one study, college students in a randomly selected group were told that a personality test showed they were likely to end life alone, while a second group was given bad news of a different kind: they were told that the test showed they were likely to have many accidents later in life. The first group performed much worse on an IQ test or the analytical section of the Graduate Record Exam (GRE) immediately after the prediction. They also had more trouble recalling information for a difficult reading comprehension task. The effects were large, corresponding to a twenty-five-point drop in IQ.

Social rejection did not affect performance on less-demanding tasks. The two groups showed no difference in their ability to memorize nonsense syllables or their ability to answer easy reading comprehension questions. Based on these results, the researchers suggested that the prediction of future social rejection impaired the ability to reason because it depleted participants' capacity for self-control (see chapter 13), a finding that was later confirmed.

Parents concerned about academic achievement might do well to focus on building their children's self-control ability and social skills (chapter 20). There is good evidence that these capabilities can be modified by experience, and they contribute not only to a happy and successful life but also to intellectual achievement.

The question of how much circumstances can modify people's intelligence remains controversial even among academics. This argument is ongoing partly because the development of intelligence has political implications for disadvantaged groups and partly because the issue is genuinely complicated. You may find this chapter easier to follow if you first go back and read chapter 4 and the box in chapter 10 (*Practical tip: Outdoor play improves vision*, p. 84) to remind yourself of how a trait can be highly heritable and strongly affected by the environment at the same time.

Let's start by defining what we mean by *intelligence*. Individual people's performance on any cognitive test is moderately predictive of their performance on any other cognitive test. These broad correlations between different cognitive skills reflect the existence of a general reasoning ability, often called *g*, which is measured (though not perfectly) by IQ tests. Intelligence can be subdivided into knowledge (*crystallized intelligence*) and reasoning skills (*fluid intelligence*).

People with better fluid reasoning skills process information more efficiently, reacting more quickly to stimuli and requiring less brain activity to solve problems. Intelligence is closely related to working memory, the ability to hold information in your mind temporarily while you're doing something with it. Working memory can be as simple as remembering how much salt a recipe requires while you reach into the cupboard, or it can be as complicated as keeping track of the steps in a multistage process while you're also evaluating whether it's working correctly. People with good fluid reasoning ability are resistant to distraction; they are less likely to lose their place when they return to a task after temporarily turning their attention to something else.

During development, fluid reasoning ability first emerges at age two or three. It grows quickly until middle childhood and then more slowly until it reaches a plateau in midadolescence. After that, it declines very slowly through adult life and then more quickly in old age. (In contrast, crystallized intelligence continues to increase through adulthood and remains stable in old age, except in cases of dementia.)

There are many uncertainties involved in measuring (or perhaps defining) intelligence early in life, before it has fully developed. One of the best early predictors of IQ is habituation in infancy, that is, how quickly a baby becomes bored with a new stimulus and looks away. This measure is not very reliable, as it predicts only 17 percent of the variability in later intelligence and often produces different results when researchers retest the same child over time. Cognitive testing in babies and young children can distinguish mental retardation from normal intelligence but cannot distinguish moderate differences in normal intelligence. The results of IQ tests get progressively more stable through childhood, becoming fairly reliable around age seven or eight, and settling around the future adult score by age twelve. In other words, young children's brains are not finished enough to allow parents or teachers to determine who will turn out to be merely bright and who will be truly gifted.

Intelligence is a strong predictor of later academic and professional achievement, social mobility, and even physical health. Still, parents (and teachers and everyone else) should keep in mind that intelligence accounts for a bit less than half of the variation among individuals on most cognitive tests. The remainder of the test variance is attributable to mood, motivation, specific cognitive strengths and weaknesses, and experience with the particular test and with testing in general. Self-control too is important for later achievement; as we pointed out in chapter 13, the marshmallow test at age four predicts later SAT scores twice as well as IQ. Finally, life success depends not only on ability but also on opportunity and effort.

Twin and adoption studies demonstrate conclusively that genes can strongly influence individual differences in intelligence. Among middle-class populations, the heritability of intelligence increases substantially with age, from 30 percent in early childhood to 70–80 percent in late adolescence and adulthood. This change occurs partly because as people grow older, they become more able to choose environments that suit their personal characteristics. In particular, intelligent children tend to place themselves in intellectually stimulating circumstances if they can, which improves their cognitive development. This gene-environment interaction increases the apparent strength of genetic influences (see chapter 4).

As we noted in that chapter, the effect of genes appears far weaker under conditions of deprivation. If the environment is sufficiently impoverished, there is little or no correlation between genetic inheritance and intelligence. This probably happens because such environments offer children few opportunities to develop their genetic potential. Thus improving the environment may paradoxically increase the apparent contribution of genetic factors to individual differences. Similarly, a meta-analysis shows that adopted children have higher IQs, on average, than their siblings who remained in the birth family, presumably because adoptive families with higher socioeconomic status provide an environment better suited to cognitive development (see chapter 30).

Curiously, although researchers have identified about three hundred genes associated with mental retardation, it has proven difficult to link particular genes to normal variations in intelligence. One reason, of course, is that intelligence is influenced by many genes—which makes isolating them that much harder.

For comparison, forty genes are known so far that contribute to height, but they explain a total of only 5 percent of variation among individuals; more than likely, many more height genes remain to be identified. The genetics of intelligence seems likely to be even more complicated.

Brain structure itself is heritable and is correlated with intelligence. This correlation becomes stronger with age, and both measures are influenced by similar genes. Overall brain size moderately predicts intelligence (though there are exceptions, most notably that men have larger brains than women but equal intelligence), as do the volumes and cortical thickness of various brain regions. In a longitudinal study of children, the pattern of developmental changes in cortical thickness predicted intelligence more strongly than did the adult configuration at age twenty (see chapter 9). Dendritic branching in neurons was also correlated with intelligence in a few studies.

Parents can encourage their children to handle failure constructively by praising them for what they do rather than for what they are.

As you might imagine, intelligence is not located in a single brain region. Reasoning requires the transfer of information between brain regions, so brain connectivity is important. Among the major bottlenecks are certain long-distance connections within the brain; intelligence is correlated with making these connections easily and effectively. One study found that links from a region of the prefrontal cortex to the anterior cingulate and parietal cortex were stronger in more intelligent children and adolescents.

In adulthood, a network of frontal and parietal cortex regions seems to be most important for intelligence. In patients with brain damage, lesions in the left frontal and parietal cortex impair working memory, a key component of intelligence. Damage to the left inferior frontal cortex impairs verbal comprehension, and right parietal damage impairs perceptual organization. An area called the *rostrolateral prefrontal cortex* seems to be activated specifically during tasks that require you to integrate multiple mental representations, as when plotting your next move in a game of chess.

A handful of imaging studies suggest that children may use their brains differently from adults during abstract reasoning. Compared to adults, children ages six to thirteen recruit the rostrolateral prefrontal cortex for easier tasks. Children use this brain area to answer single-relationship questions like "What is the best match to a fish? (a) field, (b) water, (c) tree, (d) oatmeal," while adults activate it only for two-relationship questions, such as verbal analogies ("Leaf is to tree as petal is to what?").

The role of the environment in the development of intelligence is also substantial, but it has proven more difficult to define, in part because of the gene-environment interaction discussed earlier in this chapter. James Flynn—a moral philosopher who turned to social science and statistical analysis to explore his ideas—was the first to note one strong environmental effect on intelligence. Average IQ scores have risen recently by three points per decade in many countries, and even faster in some countries, such as the Netherlands and Israel. For instance, in verbal and performance IQ, an average Dutch eighteen-year-old in 1982 scored twenty points higher than the average person of the same age in his parents' generation in 1952. The same improvement was seen when these eighteen-year-olds were compared with their own fathers at a similar age. By now the existence of the *Flynn effect* has been established beyond any reasonable doubt.

It might be tempting to say that this younger generation had evolved to be more intelligent than their parents, but evolution takes much longer than a few decades. Flynn points out that modern times have increasingly rewarded complex and abstract reasoning, and that this has been happening for over a century. This environmental change may be responsible for increasing IQ over time, both directly (by increasing your child's reasoning ability) and indirectly (by increasing the reasoning ability of others in your child's social group and thus making the social environment more complex). The change appears to be restricted to fluid intelligence, since capacities requiring less of it, such as vocabulary or arithmetic, have not shown comparable increases.

So if changes to the world can make your kids more intelligent, shouldn't we be able to control that as parents? Well, yes and no. Some experimental interventions can increase intelligence—in children and even in adults. But there's a catch: the successful approaches all require a lot of hard work in exchange for a small increase in reasoning ability. There are good reasons for your child to devote months to learning to play a musical instrument, but a gain of three IQ points isn't the first among them (see *Practical tip: The benefits of music and drama*, p. 200). Similarly, in adults, extensive practice on a difficult working memory task leads to a four-point IQ gain. Researchers do not yet know whether such gains last after you stop training or whether they translate to improved professional or academic performance—or any other desirable outcome. These studies are promising, but they are not yet ready for widespread application.

You have probably seen a variety of advertisements for products that claim to increase your child's brainpower, but we are skeptical of their value. The marketing departments have gotten far ahead of the data in claiming gains in brain function from programs that have not been tested adequately—or in most cases, at all. Some of these programs are actively detrimental to children's development (see *Practical tip: Baby videos do more harm than good*, p. 142). Even those that do no damage directly are displacing other activities that may benefit children more, such as free play (see *Practical tip: Imaginary friends, real skills*, p. 117) or time spent outdoors (see *Practical tip: Outdoor play improves vision*, p. 84).

In all, we suspect that unless your children enjoy such activities for other reasons, their time may be best spent discovering the pursuits that motivate them to achieve excellence for their own satisfaction. Childhood offers the chance to develop a variety of abilities that are important to a well-balanced life. Once your child has found an activity that matches his interests and abilities, you may find that effort and opportunity take care of themselves.

Chapter 23

TAKE IT FROM THE TOP: MUSIC

AGES: BIRTH TO NINE YEARS

Significant parts of this book were written to the soaring accompaniment of Sam's three-year-old daughter belting out the songs of the ABBA musical *Mamma Mia*, over and over. When your child makes music, you've probably noticed that he seems able to stay focused for a long time—and that he finds it great fun. Whether it's singing a favorite tune or banging away on an instrument, there's something about music that can keep his attention for as long as half an hour, an eternity for a small child (and perhaps those listening to him).

Parents often try to deepen this relationship through lessons. The goals are not just aesthetic but practical: in coaxing their children to become more involved with music, many parents hope to improve their children's minds. For this reason, when Sam was a child, his parents attempted to introduce him to the accordion, which thankfully did not stick, then to the violin, and finally to the piano, which did. Products like Brainy Baby purport to make infants smarter simply by exposing them to music at an early age.

Do these interventions do any real good? Based on research, the answer is mixed. Listening to music does not make children any smarter, but it does improve their moods, which leads to some secondary benefits. In contrast, there is some evidence for direct cognitive benefits from learning to play an instrument. In both cases, though, the research literature has to be taken with a grain of salt (see *Practical tip: The benefits of music and drama*, p. 200).

The brain responds to music starting early in life, and musical aptitude continues to develop through age nine. As we discussed in chapter 2, the brain develops from back to front, with the brainstem maturing first, followed by midbrain structures, and finally the neocortex. This general sequence is reflected in the order of development of a child's capacities for recognizing music.

Music perception emerges not long after birth and becomes apparent during the first year of life. From the start, infants prefer higher-pitched singing to lower-pitched singing, as well as a song that is specifically directed toward them or sung in a loving tone. Babies can also perceive complex sounds such as a piano note, which combines multiple frequencies.

Infants even have innate ideas about what notes go together. One example is *consonance*, which occurs when one note occurs at a frequency that is at a simple multiple of another (for example, frequency ratios of 2 to 1, a perfect octave; or 3 to 2, a major fifth). You'll find consonance in most songs, including simple children's songs like "Twinkle, Twinkle Little Star." *Dissonance* is used for effect in music but also occurs in abundance when your child bangs a fist on the piano. Infants look longer toward the source of a note when it is followed by a consonant note than when it is followed by a dissonant note such as an augmented fourth (F-sharp compared with C, a ratio of 45 to 32)—blech! This preference is apparent as young as two months of age. Indeed, dissonance is such a turnoff that when it occurs at the beginning of the experiment, infants check out permanently and stop looking at the speakers.

MYTH: THE MOZART EFFECT

The belief that passive experience leads to brain improvements is widespread. One of the most persistent brain myths is that playing classical music to babies increases their intelligence. There's no scientific evidence for this idea, but sellers of classical music for children encourage the belief every chance they get.

This myth began with a 1993 report in the scientific journal *Nature* that listening to a Mozart sonata immediately beforehand improved the performance of college students on a complex spatial reasoning task. The researchers summarized the effect as equivalent to an eight- to nine-point gain on the Stanford–Binet IQ scale. Journalists were not especially interested in this finding when it was first published.

The turning point in this idea's popularity was the publication of *The Mozart Effect* by Don Campbell, an influential bestseller. A low point of the craze was reached when Georgia governor Zell Miller played Beethoven's "Ode to Joy" to the legislature and successfully persuaded them to spend $105,000 to send classical music CDs to all parents of newborns in the state. Almost two decades later, the idea that classical music makes babies smarter has been repeated countless times in newspapers, magazines, and books. It is familiar to people in dozens of countries. In the retelling, stories about the Mozart effect have progressively replaced college students with children or babies. Some journalists assume that the work on college students applies to babies, but others are simply unaware of the original research.

Since 1993, scientists have attempted to repeat the original experiment on college students, with mixed success. The closest that anyone has come to testing the idea on babies is that preschoolers do somewhat better on cognitive tests after hearing age-appropriate children's songs. Even then, like the college students listening to Mozart, the effect is short-lived, not long-lasting, and probably attributable to improvements in their mood.

Beyond consonance, infants are open to different tonal structures. Eight-month-olds are equally good at distinguishing changes in Western scales or Indonesian scales, which are quite different once you get past the basic octave-level similarity. This ability arises around the same time that infants develop the ability to distinguish vowels and consonants of their own language from other languages (see chapter 6). Western adults and older children, however, are much better with Western scales. Infants are similarly open to different culture-specific rhythmic structures. Musical capacities are another example of how the brain's abilities and preferences are tuned to match the local environment during development.

Babies absorb major features of music, like language, well before they can produce it. One example is rhythm and meter. In Kindermusik, a method of early childhood education in music and movement, infants are exposed to rhythm by activities such as being bounced on the knee in time to a simple, repetitive beat. After two minutes of bouncing to every other beat, the baby is more interested in new rhythms of that type, as opposed to rhythms that emphasize every third beat in a waltzlike fashion. The same effect happens in reverse if babies are bounced in a waltz rhythm. So movement influences auditory rhythm perception in infants, and classes such as Kindermusik may enhance the development of culture-specific rhythm preferences.

In the preschool years, additional capacities develop. By age four, children show a good ability to detect different intensities (loudness), followed by frequencies (pitch), and finally tone duration. Intensity and frequency are processed in the earlier-maturing lower auditory system, while duration requires later-maturing structures such as the neocortex. This developmental process depends on experience. In deaf children who receive cochlear implants, the process is delayed by the period of deprivation, even though their rhythm processing is normal. The need for learning may explain why young children have difficulty staying in tune or on rhythm.

Around the same time, children develop an advanced aspect of music processing: key and harmony. As early as age three, children know whether notes are in key, can pick out dissonant notes in a familiar song, and even adjust their pitch to match another singer. At this age, children also can detect harmony between notes played together, an ability that emerges clearly by age six. Key and harmony preference are both refined by music training. Progress in these areas adds up to general musical aptitude, which by age nine reaches a degree of maturity that remains the same throughout life. By then, parents and teachers can get a sense of what kind of musician a child could become. If by that age she's got a tin ear, perhaps it's time to reconsider those flute lessons.

As you might expect, many brain structures that reflect musical aptitude have an auditory function. The auditory cortex, which is found in the temporal lobe, below the temporal-parietal sulcus, is a major site of music processing. The auditory cortex is largely found in structures called *Heschl's* **gyrus** and the *planum temporale*, whose sizes stabilize around age seven. The hemispheres seem to be specialized, with a note's fundamental frequency processed in the left hemisphere and its spectral pitch (the actual frequencies contained in the note) in the right hemisphere.

In adults, the size of these structures is strongly related to musical ability—the largest known structural variation connected with ability in people. Trained musicians, whether professional or amateur, have over twice as much gray matter in anteromedial Heschl's gyrus compared with nonmusicians. The additional gray matter is active, too; when a tone is played to musicians, they produce characteristic brain signals, found between fifteen and fifty milliseconds after the tone, that are considerably larger than in nonmusicians. At fifty milliseconds, the signal is five times as large in musicians.

These distinctive characteristics suggest that it might be possible to predict musical aptitude based on brain anatomy. To an extent, this is true: the gray

PRACTICAL TIP: THE BENEFITS OF MUSIC AND DRAMA

Playing classical music for your kids isn't likely to help their brain development. But what about having them play music for you?

Music lessons certainly make children better at music-related capacities. These improvements can begin as early as age three and continue as children advance musically. A more common question, though, is whether music training makes a child smarter. Psychology journals are filled with studies showing that music lessons predict visual, motor, attention, and mathematical skills. There is one difficulty with many of these studies, though: nearly all of them are correlative. Perhaps the musically experienced people who took these surveys were smarter to begin with. They often come from advantaged families, as parents who fund music lessons tend to be well educated and financially secure.

The psychologist E. Glenn Schellenberg sought to eliminate this vari-

able. He placed advertisements, recruiting families with six-year-olds for free art lessons. They were then split into four groups: standard keyboard lessons, age-appropriate vocal music lessons, drama lessons, or placement on a waiting list for one year (after which they received the promised art lessons). The drama and lessons-deferred kids provided two control groups against which the other two groups could be compared. Music lessons were given at Toronto's prestigious Royal Conservatory of Music.

The children were given IQ tests before lessons began and then tested again after thirty-six weeks. On average, the children receiving music lessons scored not quite three IQ points higher than the two control groups, which were similar to each other. The differences were spread over categories that included resistance to distractibility, processing speed, verbal comprehension, and mathematical computation. The effect size was modest, $d' = 0.35$ (see p. 64). That is, the average child receiving music lessons scored better on the IQ test than 62 percent of the control children.

Drama classes showed an unexpected, larger benefit: better social adaptation. Children who took drama lessons at the conservatory showed marked improvement on a rating scale for adaptability and other social skills. The effect size was $d' = 0.57$, so that the average child receiving drama lessons scored higher than 72 percent of children in the nondrama groups. This effect is large enough that you would be likely to notice the change in your own child. Perhaps deliberate practice at inhabiting the character of another person served to improve the performance of brain areas involved in daily social interactions.

In general, practicing any activity is likely to have the strongest effect on the brain capacities that the activity directly requires. Learning to play music does have some ancillary benefits that span a variety of cognitive abilities, perhaps because it trains the brain's attentional networks or because musical performance calls so many brain regions into action. In the end, the benefits appear to be small—but real.

Compare this with the intrinsic benefits of music for its own sake, such as whether your child likes playing the instrument or will enjoy music more in adulthood. This benefit of musical training goes beyond the utilitarian. Training gives your child access to music at a deeper level and can contribute to a lifelong love of music.

matter volume of anteromedial Heschl's gyrus is related to musical aptitude with a correlation coefficient (called *r*) of about 0.7. This translates to the statement that if an adult has an above-average amount of gray matter in this brain region, odds are about 3 to 1 that he or she is above average in musical aptitude.

Does experience influence the size of this brain structure? A clue that such a change is possible can be found in the white matter, which contains long-distance connections between distant brain regions. The long-distance connections through the white matter are organized differently in professional musicians than in nonmusicians. If musical training starts earlier in childhood or involves more cumulative hours of practice, these differences are larger. So extended childhood practice seems to result in measurable changes along with the hard-won skills, though it remains possible that children who started with a larger Heschl's gyrus were more likely to stick with music lessons.

It might be possible to predict musical aptitude based on brain anatomy.

More convincing evidence comes from a prospective study that followed two groups of children for fifteen months in which one group took weekly keyboard lessons, while the other group participated in a school music class involving singing while playing rhythm instruments. There were no differences in brain structure at the beginning of the study, when the children were six years old on average, but by the end, the keyboard group had larger volumes in the frontal gyrus and the **corpus callosum**. Increases in the size of the corpus callosum, which links the halves of the neocortex, are likely to lead to faster communication between the two hemispheres, which should facilitate the production of well-coordinated two-handed movements. For children who later became musicians, the difference in corpus callosum size persisted into adulthood. There was also some increase in the size of Heschl's gyrus, but it did not reach statistical significance. Longer follow-up may show a practice-based increase in this region as well.

The processing of melodies involves additional brain regions, including the temporal and frontal areas of the neocortex. These regions are critical for tonal working memory, for instance, holding a melody in your head.

Playing music calls into action yet more brain regions, as the musician must generate precisely timed sequences of motor activity. In brain scanning experiments, activity during sequence learning and production encompasses motor-related regions of the neocortex as well as the basal ganglia and the cerebellum of musicians and nonmusicians alike. These brain structures are necessary for the initiation and guidance of movement. In the case of music, the demands of coordinating auditory experience and fine movement are particularly intense. How this coordination is achieved by the brain is a very active area of research.

The same brain regions, as well as parietal parts of the neocortex, are activated when people listen to musical sequences. The cerebellum is active in both musicians and music listeners, which likely means it is involved in both producing precise timing and processing purely auditory information. Even more brain regions are activated when people hear complex sequences and combinations of notes.

Such widespread recruitment of brain areas is not surprising if we think of the complexity of recalling a musical sequence. Try to remember the following numerical sequence: 1, 1, 3, 5, 8, 6, 6, 4, 5, 6, 5. Doable, certainly, but you'd have to rehearse it in your head many times to get it. But what if those numbers were translated into music notes, like this?

Even a nonmusician can memorize this melody. Add rhythm and harmony, as you probably did without trying if you know the song, and the piece becomes even more complex. A powerful property of music is its capacity to call into action mechanisms for recalling and producing sequences with a rich structure. In this regard, students of music achieve levels of memorization, recall, and technique that are aided tremendously by music's capacity to guide and organize brain activity. Music provides scaffolding for mental feats that are otherwise hard to attain.

Chapter 24

GO FIGURE: LEARNING ABOUT MATH

AGES: BIRTH TO EARLY TWENTIES

As Barbie famously said, "Math class is tough!"—and not just for girls but for everyone. Your child's brain is optimized to provide rapid solutions to everyday problems. That means it is less suited to solving an algebraic equation than to calculating whether it would be a good idea to punch that kid who just insulted him. (Of course this social calculation does involve some numerical ability, since it's important for your child to determine whether the other kid has more friends available nearby than he does.)

Young babies and many other animals share a brain system that supports this sort of rough number sense. Under the right conditions, this number sense can combine with our species' ability to create and manipulate symbols to produce formal mathematics, found in some societies but not others. Indeed, math, a seemingly inhospitable place for dandelions to grow, is surprisingly fertile ground.

In the last few decades, our appreciation for babies' ability to form number-related concepts has expanded tremendously. Infants express surprise by looking longer (see chapter 1) if one object goes behind a screen and two come out. If an infant sees a Mickey Mouse doll go behind a screen and then the screen lifts to show a truck, she doesn't care. If she sees a Mickey emerge along with a second Mickey, now that's a surprise, as evidenced by her long gaze. This ability to notice an extra object—the twoness of the Mickeys—is a necessary component of numerical concepts.

This ability goes beyond small numbers. When a six-month-old infant sees a

series of pictures, each containing a number of objects—dots, faces, anything—he will notice if the number either doubles or decreases by half. This general sense of *numerosity* gets better with age, too. While infants can recognize a 1:2 ratio (for instance, comparing 4 and 8 objects, or comparing 6 and 12) without counting, adults can recognize a more subtle 7:8 ratio.

Numerosity detection, the ability to distinguish between groups of different sizes, is an ability that all humans have. *Subitization*, another universal capacity, refers to the ability to immediately distinguish small numbers without counting. The term comes from the Latin word *subitus*, which means "sudden." Both numerosity detection and subitization are apparent in other animals—and they involve some of the same brain mechanisms in people.

These abilities, which can be seen in mice, dogs, and even pigeons, provide an obvious survival advantage: they allow us to estimate the quantity of something, from food sources to possible enemies. For instance, a subgroup of a pride of lions reacts differently to roaring sounds depending on how many lions they hear—and on how many members are in their own group. If they are outnumbered, they call to the rest of the pride for backup. Similarly, chimpanzees avoid conflict with other groups when they are outnumbered.

One reason it took so long to understand young children's number sense is that early researchers (such as Piaget) asked the wrong questions. If asked "which row has more objects?" children aged three or four will point to a smaller number of clay pellets if that row is spaced out to look longer. Change the pellets to chocolate candies that they can have right away, though, and children do much better. In retrospect, it appears that this research tested for two things: a sense of number and the ability to express it in a clear way. Your three-year-old knows, but evidently she isn't saying. Aside from her mouth being too full of chocolate to talk, her awareness isn't accessible to an interviewer's questions.

Oddly, two-year-olds do fine with either pellets or candies. This result seems to imply that at that age, children have a clear sense of numerosity, but then they lose the abstract sense of it for about a year. What could be happening here? One possibility is that at three or four, children's brains are in the midst of hooking up their intuitive understanding of quantity with an explicit, later-developing sense of abstract numbers. By five it's all sorted out, at which point she simply counts the pellets—and perhaps wishes for candies instead.

Grabbing candy bits may seem somewhat primal, as indeed it is. Evidence suggests that chimpanzees can also combine quantities in a mental operation that resembles addition. If a chimpanzee is shown two trays in succession, each with a different number of chocolate bits, he can determine whether the combination of the two trays contains more or less than a third tray. The rudiments of arithmetic are therefore evolutionarily older than our species and are basically one facet of your toddler's inner ape.

These senses of number involve similar brain regions in people and chimpanzees. Numerical information seems to be represented in the prefrontal and posterior parietal lobes. One key location is the *intraparietal sulcus*, a buried groove in the brain where specific semantic number content (for example, *seventeen*) is represented. When this brain region is damaged, people can give approximate but not exact answers—at about the level of a chimpanzee.

This retained ability for general numerosity has led scientists to suggest that our brains represent numbers in a way that relates to their relative magnitude, as a mental number line. One piece of supporting evidence for this idea is that when we are asked to judge which of two numbers is larger, it takes longer to answer when the two numbers are close (8 versus 7) than when they are far (8 versus 2), as if the closer numbers are actually closer in mental space. Judging between closer numbers produces more activity in the intraparietal junction. You could imagine numbers being stored in some computerlike, digital fashion, where small differences are just as easy to detect as large differences, but instead, brains seem to use a more ordered representation, akin to marks on a ruler.

Can an individual child's general capacity to handle quantity be trained even before she begins to count?

In monkeys, some neurons of the left and right intraparietal sulcus fire when the animal encounters a particular number of objects—or an approximately similar quantity. In general, these brain regions are part of the same major pathway in the brain for identifying where things are, including how many things are there and where they are going.

The "where" capacity of the parietal cortex (see chapter 10) seems to encompass a variety of functions. The posterior parietal cortex becomes active, in both monkeys and people, in conjunction with eye movements. In relation to math, neuroscientists noticed a curious additional capacity of this brain area by having people do simple math while they were lying in a fMRI scanner: the same regions become active when people do mental addition and subtraction problems, even though the eyes are not moving. Nearby parts of the brain with many shared connections to this region are intimately involved in visual functions, such as the abrupt eye movements called saccades, attention, and detecting which way a visual pattern is moving. Thus the way we look at space might be closely tied up with our mental number line. The pattern of activity in the posterior parietal cortex can even be used to predict with middling accuracy whether a person is adding or subtracting.

This seemingly odd overlap in the brain of eye movement commands and basic arithmetic suggests that some aspects of our brain's ability to process

PRACTICAL TIP: STEREOTYPES AND TEST PERFORMANCE

People's performance changes a lot if they're reminded of a stereotype just before an exam—even by checking a box for male or female. Any relevant negative stereotype can impair performance, especially when people believe that the test is designed to reveal differences between groups. Stereotypes can be activated even if test takers are not aware of the reminder, for instance, when African American faces are briefly flashed on a computer screen. Even more curiously, these effects can occur in people who are not members of the stereotyped group: young people walk more slowly after hearing stereotypes about the elderly. This appears to happen because thinking about the stereotype takes up working memory resources that would otherwise be used for the test.

A little effort can minimize this problem. Obviously, teachers shouldn't expect certain students to perform poorly. Standardized tests should collect demographic information at the end of the answer sheet. The effect also works in reverse: girls do better on a math test after hearing a lecture on famous female mathematicians.

Most people belong to more than one group, so perhaps the most practical approach is to bring a more positive stereotype to the task. For example, a mental rotation task shows consistent sex differences, with men faster and more accurate than women. When reminded of their gender before this test, women got only 64 percent as many correct answers as men. In contrast, when reminded that they were students at a private college, the women got 86 percent as many correct answers as men. The men did better when reminded of their gender, while the women did better when reminded that they were elite students. Thus the gap between men's and women's scores was only a third as large when women were reminded of a positive stereotype that fit them as opposed to a negative stereotype. The remaining gap is likely to be a real sex difference: a single shot of testosterone temporarily improves women's performance on this test.

Stereotyping is a strong brain tendency, unlikely to disappear soon (see *Did you know? Stereotyping and socialization*, p. 174). Instead, we recommend taking advantage of such brain shortcuts by reminding your children of a stereotype that will improve their performance.

abstraction are built upon our capacities for dealing with the physical world. Many cognitive abilities other than arithmetic seem to be "embodied" in a similar manner. In this way, we are able to think abstractly with brains that evolved for more concrete actions, such as looking for prey or finding a path through a forest.

Converting these approximation abilities into the precise representations of formal mathematics requires symbolic representation. This capacity comes with language, which is an elaborate means of representing information efficiently. Parrots, dolphins, macaque monkeys, and chimpanzees can use symbols to represent numbers. For example, two macaques named Abel and Baker were able to pick the larger of two digits to get a larger number of candies. For the most part, animals cannot combine the symbols to add or subtract. One exception is a chimpanzee named Sheba, who after several years of training could perform some simple addition.

Even though people have the mental capacity for arithmetic and mathematics, they don't always use it. The researchers Stanislas Dehaene and Pierre Pica investigated the Mundurukú, an Amazonian group that lacks arithmetic and has very few words for numbers. A few of the words they do have are precise (*pug ma* = one, *xep xep* = two), but most are approximate (*ebapug* = between three and five, *ebadipdip* = between three and seven). Mundurukú do approximate addition of large groups of objects very well, performing as accurately as Western, numerate adults. But exact calculation of small numbers is beyond them; for instance, if six beans are placed in a jar and four are drawn out, when asked how many are left, they say "zero" or "one" more often than they say "two."

Formal arithmetic ability is predicted by a child's earlier capacity for approximate number. This suggests that individual children differ in their general ability to handle quantity even before they begin to count. Can this capacity be trained? Perhaps children could be taught to do approximate number problems to improve their later acquisition of arithmetic skills. Although this idea has not been tested, it presents an intriguing possibility.

From the basic senses of number—subitization, numerosity, and symbolic representation—we can construct a host of more complex concepts, such as negative, fractional, and real numbers. From these and additional brain capacities, it becomes possible to imagine the universe of mathematics: multiplication, trigonometry, functions, calculus, and more.

The study of how the brain produces abstract mathematics has barely begun, but researchers have taken a few first steps. At higher levels of math, additional concepts—and many brain regions—come into play. Algebra requires kids to combine their numerical abilities with symbolic, abstract manipulation. Beginning students can come at algebra by different routes. For instance, it is often easier to solve a word problem than to solve an equation. These different approaches emphasize different brain regions.

To monitor what brain regions are activated by different approaches to solving a problem, researchers took fMRI scans of people doing story problems (If Cathy makes $10 an hour and gets $12 in tips at the end of a four-hour shift, how much money has she earned in total?) and similar equation problems (If $10H + 12 = E$ and $H = 4$, what is E?). The scans showed that solving story problems preferentially activates the left prefrontal cortex, an area associated with working memory and quantitative processing. Equation problems activate regions associated with the mental number line, such as parts of the parietal cortex, including the *precuneus* (a portion of the parietal lobe facing the midline), as well as parts of the basal ganglia that are essential in nonalgebraic life for action and movement.

This difference suggests that beginners at algebra may want to try several different approaches to the same problem. For harder problems, in addition to the cortical areas we mentioned, many more regions in the left hemisphere are activated. Higher math such as trigonometry or calculus has not been investigated thoroughly, but researchers believe that these capacities also build upon brain systems for symbolic and spatial manipulation.

At some level, this supports Euclid's aphorism about geometry that "there is no royal road to learning." Mathematics is an incredibly complex system, one of humanity's great achievements, and it's remarkable to think that brain circuits for telling stories and moving the eyes have been harnessed to generate, understand, and use it. It's a feat of matching the brain to an environment that our ancestors never imagined.

Chapter 25

THE MANY ROADS TO READING

AGES: FOUR YEARS TO TWELVE YEARS

The human brain took its present form before the first word was ever written. In the five thousand years since the alphabet was created, our brains haven't changed that much. So reading (like advanced math) must use circuitry that originally evolved to fulfill other functions. Brain scanning studies have started to reveal how the systems for reading mature. In children who learn alphabet-based languages such as English, patterns of brain activation change in a sequence that reflects certain stages of development. These steps follow a different trajectory in dyslexic children—but also in children learning to read Chinese. Evidently the road to literacy takes multiple paths, some of which may be smoother for one child than for another. The right choice of a language may improve the odds of a good outcome.

At its core, reading consists of learning the relationship between words and marks on paper. In Western languages, the marks are letters; in Chinese, characters. Most children start learning to read and write around age five or six. Over years of practice, this process becomes automatic and effortless.

As neuroscientists have found by using brain scanning technology on adults, during reading the brain shows activity in many regions. These include the frontal lobe, the cerebellum, and the area where the temporal lobe of the neocortex meets the parietal and occipital lobes. One especially important region is a part of the fusiform gyrus within the left inferior temporal cortex. This region appears to have a special importance in the recognition of written language (which is

why it's sometimes called the *visual word form area*). This region is active when a person is shown either words (*cat*) or groups of letters that look like words but are not (*zat*). This discrimination is not simply a matter of recognizing whether the pattern is pronounceable. Chinese characters are composed of stroke patterns with no intrinsic pronunciation—you can't "sound out" the pronunciation of a Chinese word; you just have to know it. And yet nonsense characters that resemble real words also activate the visual word form area in Chinese readers.

Word-form recognition is learned through experience. This capacity seems to be an example of a more general ability of the inferior temporal cortex to visually recognize objects (see chapter 10). Some neurons fire when a monkey or person sees specific objects, such as a flower, a hand, or a face of a monkey or person. Neural processing for faces is concentrated in the right hemisphere. "Face neurons" in the inferior temporal cortex are quite specific, responding only to faces and not to individual features, rearranged images of faces, or even upside-down faces. Recognition is more likely if an object is familiar or important to the viewer. For example, part of the region that is specialized for face recognition also responds to specific car models—in the brain of a car expert.

Most naturally occurring objects look the same when viewed from the left or the right. Perhaps for this reason, mirror-image confusion is common in animals and people. Your child's brain, and yours too, are optimized to solve common visual problems—but for most of evolution, those problems have not included reading. There may not be much advantage to distinguishing left- from right-side views of objects most of the time. Therefore the right inferior temporal cortex may not be wired to detect asymmetry. This characteristic suggests why it might be useful for the inferior temporal cortex on this side of the brain to drop out of the reading circuit.

Attempts by your child's right inferior temporal cortex to participate in reading can pose a problem, because words and letters are loaded with asymmetry. In reading, the ability to distinguish mirror images is often important, for instance, in distinguishing *b* from *d* or *AM* from *MA*. As a result, the brain must suppress any tendency to perceive left and right views as being the same, which may explain why many brains don't take naturally to reading. Mirror-image confusion might interfere with recognition of letters and words—and in some cases be involved in difficulties with early reading.

Because natural objects such as faces look the same from either side, the more efficient recognition strategy is often to treat left and right views as being the same. Overcoming the natural tendency to treat mirror images as equivalent is a major milestone for beginning readers. In kindergarteners, the ability to make left-right distinctions is correlated with readiness to read, and children at this age routinely reverse letters. The relationship between left-right detection capacity and early reading disappears by first grade, suggesting that most children clear this hurdle by the age of six. From then on, readers rely more and more on regions in the left frontal and temporal lobes. Conversely, dyslexics frequently have left-right confusion, as well as difficulty distinguishing mirror images from each other. Perhaps for this reason, dyslexic children often retain a capacity for mirror writing, which most children lose.

Even though monkeys can't read, they share with your child a natural affinity for visual symmetry. Studies in monkeys suggest that suppressing that affinity is likely to come with decoupling of the right inferior temporal cortex. Like most animals, monkeys have trouble telling apart left-right asymmetric stimuli. After damage to their inferior temporal cortex, they do not get worse at this task—and sometimes they get better. It seems as if part of the work in this task is overcoming the natural tendencies of the inferior temporal cortex for shape recognition.

The right inferior temporal cortex may be the culprit in early reading difficulties. One group of researchers showed words to English-language readers age six and older. In beginning readers, both the left and right inferior temporal cortex are active during reading. This balance slowly fades until, by age sixteen, inferior temporal cortex activation has shifted largely to the left side. At ages in between, six- to ten-year-olds showed a wide variety of activation on the right side. Right-side activation and reading test scores were negatively correlated. That is, kids with less right inferior temporal activation were better readers.

In the five thousand years since the alphabet was created, our brains haven't changed that much. So reading (like advanced math) must use circuitry that originally evolved to fulfill other functions.

Another likely step in learning to read successfully is the ability to identify and manipulate spoken sounds, for instance, being able to judge that *bat* and *cat* end with the same two sounds. This capacity is known as *phonological awareness.* Phonological naming of letter sounds predicts reading proficiency, and deficits in this capacity may be a central cause of dyslexia. One brain region activated in early readers, the left posterior superior temporal sulcus, is more active in children with a better ability to recognize and classify spoken sounds. More broadly, in both early and mature readers, activity is seen in areas in the temporal and parietal cortex. These brain regions are well positioned to receive both auditory and visual information, and so might be important for integrating these modalities with each other for word recognition.

The involvement of brain regions that process sound makes sense in the case

PRACTICAL TIP: READING AT HOME

Many videos claim to teach early reading to infants. The creators of such products state that literacy skills are best taught from birth to about age four. Yet during this time, children lack the capacity to distinguish *b* from *d*, much less read whole words. No studies show that babies are doing anything more than forming associations when they watch these videos. To paraphrase one product: no, your baby can't read.

The timetable of perceptual development does raise another question: what is the benefit of children's books? In young children who can't read, the benefits appear to be associated with an interactive style of reading. Rather than taking a straight reading approach or even asking questions that can be answered by having the child point, you can accelerate language development through social engagement—by asking open-ended questions, questions about a character's actions or attributes, and by responding to your child's attempts to answer the questions.

Access to books is strongly linked to educational achievement. Among families with similar incomes and parental education, the number of books in the home is a good predictor of children's reading ability. On average, across multiple countries, children with many books stay in school three years longer than children without books. Having books at home makes as much difference to children's accomplishments as having university-educated rather than uneducated parents.

---

of alphabetical languages such as English, which require sounds to be linked with letters. But not all languages work this way. One example that has attracted recent interest from language researchers and neuroscientists is Chinese.

A Chinese child beginning to read confronts a formidable task. The written language is composed of thousands of different complex characters. Each character represents part or all of a word and is a dense squarish assemblage of one or more components called *radicals*. Children must learn about 620 radicals, each containing one to several dozen strokes. Finally, the visual appearance of a character usually does not reveal how it is pronounced.

DID YOU KNOW? THE CAUSES OF DYSLEXIA

Dyslexia is defined as "persistent difficulty in reading when other intellectual functions and educational opportunities are sufficient." Its frequency suggests that it is not a disorder, but simply one extreme of a range of normal variation. Indeed, in preliterate times, and even today in illiterate societies, dyslexic tendencies might not ever be noticed.

Like other neurodevelopmental disorders, dyslexia arises through genetic mechanisms. If one identical twin is dyslexic, the other twin has a nearly 70 percent chance of having the disorder too. In nonidentical twins and siblings of dyslexics, who share only half their genes, the probability is still high, 40 percent. This pattern of inheritance would be expected if dyslexia in any individual child is triggered by just one or two variant genes.

In most cases, the genes that make us susceptible to dyslexia affect either the migration of neurons to their final destinations or the growth of axons. One such gene is ROBO1, which helps to determine whether axons will cross the left-right midline. Researchers should eventually be able to understand how these genes affect dyslexia-inducing variations in neuronal circuits, either within a nucleus or in connections among brain regions.

Dyslexic children often have difficulty with phonological perception tasks, such as identifying spoken syllables or the order in which they occur. This disability might create difficulties in making rapid, automatic associations between sounds and letters, a necessary component of smooth reading. The task is difficult in English, which is filled with irregularities. Indeed, the incidence of dyslexia is lower in languages such as Italian or German, where pronunciation rules for letters are consistent—though problems still present themselves in the form of slow reading.

For some capacities other than reading, mirror confusion may be an asset. The right half of the neocortex is involved in perceptual judgments, suggesting that strong perceptual abilities in the right inferior temporal cortex might be good for other capacities. Dyslexia is frequent among artists. One survey at Göteborg University in Sweden revealed a high fraction of dyslexics among art students compared to other majors. Whether or not these students' exceptional capacities caused their dyslexia, they have identified an elite pursuit that does not require written language.

How does a child navigate such a thicket? The time-honored approach has been to learn by writing. In conjunction with a phonetic alphabetical scheme called *pinyin*, a major component of early Chinese language instruction is the writing of characters, stroke by stroke. This is very unlike reading lessons in alphabetic languages, which focus on learning individual letters and the sounds they represent. Correspondingly, there are differences among children who speak different languages in the neural pathways involved in beginning reading.

Functional brain imaging shows that, unlike English-learning children, Chinese children show only small increases in activity in the parietal cortex when reading. Instead, widespread activation occurs in a region centered on the left middle frontal cortex. The activated areas overlap strongly with the dorsal and lateral prefrontal cortex, which are used in working memory. Readers also show activity in the premotor cortex, which is likely to be activated during the execution of fine movements—such as those that would occur when writing out a Chinese character.

These findings suggest that the active recall and rehearsal of writing characters may be central to learning to read Chinese. Thus the gap between verbal language and reading can be bridged not only phonologically but also with the help of neural circuits for movement.

Phonological and movement mechanisms are independent of each other, or at least partly so. Phonological awareness is a weaker predictor of reading success in Chinese than in English. Also, the prevalence of dyslexia in China appears to be lower than in Western countries—perhaps as low as 2 percent, compared to 5 to 15 percent of English-speaking children. On the other hand, reading difficulties are more common with pinyin.

As with math, the best strategy for learning to read may differ from person to person. There are dyslexic children in China and Japan (where the written language is similar to Chinese) who reportedly learned to read English at average levels or higher by taking a phonics-based approach—but not by using a word-copying approach. For a few Chinese children, the motor-based route may be difficult. For English-language dyslexics, the phonological route may be difficult.

This brings us to another useful conclusion: if you have a dyslexic child trying to learn English, he may profit from systematic, repeated copying of entire words by hand—as if they were single symbols. Copying could activate motor circuitry to assist the mapping of language to the visual appearance of words.

This differs significantly from the usual, phonics-based approach for early reading. (It also differs from whole-language reading instruction, a rival school of reading that focuses on subject matter and context.) The seemingly laborious, frontal cortex–based approach taken by Chinese children suggests the possibility of a third route for overcoming reading difficulties: study by detailed writing.

No matter how children learn to read, the general pattern is the same: starting from a very focused group of brain regions, children eventually come to use a much broader network. The ability of the brain to organize such distant areas in the service of a cultural innovation, such as reading, is a testament to the flexibility of our brains when faced with a new opportunity.

PART SEVEN

BUMPS IN THE ROAD

Chapter 26

HANG IN THERE, BABY: STRESS AND RESILIENCE

AGES: THIRD TRIMESTER TO EIGHTEEN YEARS

Compared with adults, children start with a double disadvantage in dealing with stress: they have limited power to change their environments, and they aren't as good at managing their emotions. Every child has to find ways to deal with stress, though. It's an inevitable part of growing up, whether the problem is as ordinary as a quarrel with a friend or as serious as the death of a parent.

For adults, coping involves some combination of changing your circumstances and changing your attitude. Resilient adults are optimists. Rather than passively denying and avoiding stressful situations, they use active coping strategies such as solving the problem, reinterpreting the situation in a more positive light, seeking social support, and finding meaning in hardship. Adult resilience is influenced by early experience. In general, children seem to develop their coping skills most effectively if they are exposed to a moderate amount of stress: high enough that they notice it, but low enough that they can handle it—a level that is different for every individual and changes with age.

Stress mechanisms are similar in humans and in other animals, allowing neuroscientists to study the process in detail. Animals learn to cope with stress by starting small. Young monkeys who are separated from their mothers for one hour a week grow up to manage stress more effectively than monkeys who were never separated from their mothers. In adulthood, these mildly stressed monkeys show lower anxiety and lower baseline stress hormone levels and perform better

on a test of prefrontal cortex function. Rat pups that are separated from their mothers for fifteen minutes a day also become more resilient as adults. In contrast, pups that are separated for three hours a day grow into adults who are more vulnerable to stress, show more anxiety, are slower to learn, and drink more alcohol (when it is offered to them) than unseparated animals. The mother rat's behavior when the pups are returned to her may be one reason for the difference between these two conditions. She makes up for a brief separation by grooming the pups more, but after a long separation, she tends to neglect them.

Controllable stressors—the ones you can manage or reduce through your own actions—are more likely to lead to resilience than uncontrollable ones. Rats that learn to escape from a mild electric shock to the tail are less likely to develop learned helplessness (which psychologists consider to be an indicator of depression) when confronted with an unpredictable and uncontrollable shock later on.

Infants must rely on their parents and other caregivers to act as a surrogate coping system, as babies can signal their own needs but not meet them.

Preschoolers tend to cope in a limited number of ways, by seeking help from caregivers, confronting the problem, withdrawing, or distracting themselves with another activity. Around age three, they start to try negotiating for what they want. Older children rely most heavily on the strategies of support seeking, problem solving, escape, and distraction. Cognitive strategies for distraction (such as thinking about something pleasant) and better problem-solving ability emerge in late childhood and increase into the early twenties. Rumination and anxiety also increase during this period, though, as thinking about problems doesn't necessarily contribute to solving them.

As children get older, they learn to choose different strategies to cope with different situations, and they begin to show more personal tendencies and preferences. Some of these individual differences can be traced to their temperaments. For example, children who are quick to experience anxiety or anger (see chapter 17 and *Practical tip: Dandelion and orchid children*, p. 228) are particularly vulnerable to stress, in part because they are slow to develop self-control. Parents who are overly protective of high-reactive children may interfere with their development of coping skills.

Our biological response to stress is most effective in dealing with immediate threats to our physical well-being. When a stressful event occurs, two systems become activated. First, the sympathetic nervous system releases epinephrine and the neurotransmitter norepinephrine in less than a second, preparing your body to run or fight by directing energy to muscles, increasing the heart's blood-pumping ability, and shutting down nonessential systems. This reaction is better suited to dealing with a mugger or a bar fight than with the more common stressors of a grumpy boss or a troubled marriage.

Second, the **hypothalamic-pituitary-adrenocortical (HPA) system** (see figure opposite) works on a time scale of minutes. The hypothalamus sends corticotropin-releasing hormone (CRH) into the pituitary, which releases ß-endorphin (a natural painkiller) and corticotropin into the blood. Corticotropin then signals the adrenal cortex to release **glucocorticoid** hormones (mainly cortisol in people) into the blood. In the short term, that's a good thing, leading to changes in gene expression within the brain that help repair damage caused by the initial stress response, such as replenishing energy stores. Cortisol also increases arousal and vigilance, while inhibiting other processes, including growth, repair, reproduction, and digestion, that might divert energy from the solution of the immediate problem.

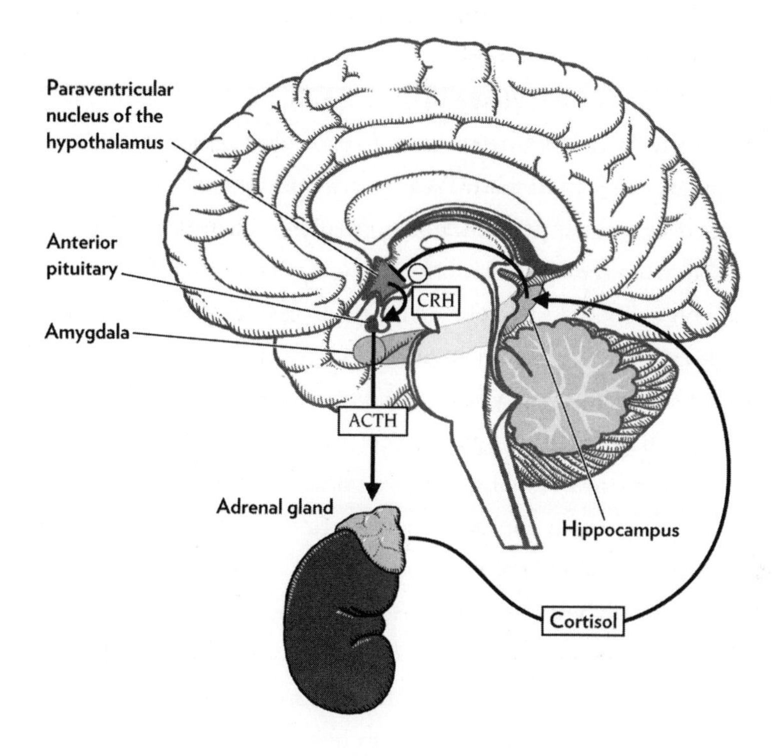

An effective stress response starts quickly and ends quickly. Binding of glucocorticoids to receptors in the hippocampus activates a negative feedback loop that shuts down CRH release and thus halts glucocorticoid production. This is important because if glucocorticoid levels stay high for a long time, you can end up with a number of ailments, including high blood pressure, damaged immune system function, osteoporosis, insulin resistance, or heart disease.

Chronically elevated glucocorticoids can also lead to brain problems. They may inhibit the birth of new neurons, disrupt neural plasticity, kill neurons in the hippocampus, or cause structural changes in the amygdala. Chronic stress makes fear conditioning easier and its extinction more difficult. Over time, the hippocampal damage not only impairs learning but also reduces the brain's ability to terminate the stress response, initiating a cycle of problems that lead to more hippocampal damage. Finally, stress causes dendrites in the prefrontal cortex to atrophy and disrupts executive function (see chapter 13).

These two systems interact extensively with each other. Neurons that produce CRH and influence stress responses are also found in the amygdala, the prefrontal, insular, and cingulate cortex, and other brain regions. The amygdala neurons, for example, project to the **locus coeruleus**, which regulates the activity of the sympathetic nervous system. In turn, neurons in the locus coeruleus can increase production of CRH in the hypothalamus.

As parents help their young children to cope with stress, the quality of their support regulates the children's HPA responses. Physiological stress responsiveness declines over the first year of life in normally developing children. They still cry to get help from their parents, but this type of crying is not accompanied by an HPA response. Toddlers and older children in secure attachment relationships show less elevation of cortisol in response to stress than children with insecure attachment relationships, those who do not see their parents as a reliable source of comfort (see chapter 20).

The strongest stress responses are found in children with disorganized attachment relationships, in which the parent often makes the child afraid, either through aggression toward the child or by being extremely anxious. Such children also have the highest risk of behavioral or emotional difficulties.

In all cases, as children become adolescents, they transition to adult patterns of stress responsiveness, becoming more vulnerable to stressful events. This may be related to the increased risk of psychopathology during those years (see chapter 9).

In most toddlers, crying is not accompanied by a stress hormone response.

The stressors that you can expect to encounter in life depend on the characteristics of your particular environment. In many mammalian species, babies' brains prepare themselves for the type of world they'll find when they grow up by tuning their stress-response systems based on their experiences during early life. This process is one of the best-understood examples of epigenetic modification, in which the environment permanently influences the way our genes are expressed and, ultimately, our long-term behavior (see *Did you know? Footprints on the genome*, p. 34).

Stress profoundly affects the developing brain. In chapter 2 (see *Practical tip: Less stress, fewer problems*, p. 16), we wrote about the effects of prenatal stress, as studied in the pregnant Louisiana women who escaped from hurricane strike zones, as well as pregnant women who experienced other forms of severe stress such as the death of a close relative. Babies born to these stressed mothers had a substantially increased risk of problems, including autism, schizophrenia, decreased IQ, and depression.

Broadly speaking, the effects of early life stress seem to be similar in people, rodents, and monkeys. Babies whose mothers were stressed or depressed during pregnancy show increased responsiveness (HPA activity) to stress later in life. Adults who were abused as children also respond to stress with stronger and more prolonged HPA and sympathetic nervous system activity than those who weren't. Probably as a consequence, previously abused adults have smaller hippocampal volumes and an increased risk of diabetes, cardiovascular disease, depression and anxiety, schizophrenia, and drug abuse. Other brain structures are likely to be affected as well, though this is less well studied.

How does early experience tune the stress response? The relationships of rats and their pups can shed some light on the subject. In rat pups, maternal licking and grooming reduce physiological stress responses. Mother rats vary naturally in how much time they spend licking and grooming their pups during the first week of life, and there is evidence that rats are able to develop normally either way. But mothers who groom a lot raise pups that are less fearful as adults and whose HPA stress responses are smaller and don't last as long, compared to animals raised by low-grooming mothers. Experiments where rat pups in one group were "adopted" by mothers from the other group confirm that this effect is due to environment, rather than genetics.

It's tempting to think of high-grooming mothers as the "good" mothers, but that would be a misconception. Instead, both types of mothers are matching their pups' adult behavior to local conditions, one of the key aims of neural development. Remember that not all stress responses are bad. Pups born into difficult circumstances may survive better if they are vigilant and their HPA system is reactive. If mothers who were high-grooming with their first litter of pups are stressed during their second pregnancy, they become low-grooming mothers for those pups and also for their third litter, meaning that their future offspring will have more reactive HPA systems.

Other indicators of a tough environment during gestation, such as protein deprivation or exposure to bacterial infection, also increase adult HPA responsiveness in rats. Not only that; pups who receive low levels of licking and grooming reach sexual maturity earlier and are more likely to become pregnant after a single mating session than pups raised by high-grooming mothers. In a tough environment (see chapter 30), these benefits for survival and early reproduction may constitute a worthwhile trade-off for a higher risk of chronic diseases late in life.

Canadian researcher Michael Meaney and his colleagues traced the molecular and neural consequences of early licking and grooming in rats. These maternal behaviors trigger the release of the neurotransmitter serotonin in the pup's hippocampus, where it initiates a series of intracellular signals that reduce the epigenetic silencing of glucocorticoid receptor genes (see *Did you know? Footprints on the genome*, p. 34). Because this DNA modification is permanent, pups that receive a lot of grooming grow up to have high levels of glucocorticoid receptors in the hippocampus and thus can terminate stress responses effectively throughout life. They also grow up to be mothers who lick and groom their own pups a lot, thereby passing the epigenetic modification to the next generation. In rats reared by low-grooming mothers, treatment with a drug that removes methyl groups from DNA can reverse the heightened HPA stress responsiveness.

The HPA stress response in monkeys is also tuned by early experience. To mature normally, young rhesus monkeys (like children) need to form an attachment to at least one reliable adult, usually the mother. Monkeys who are raised in a peer group instead of by mothers (a situation that, in humans, would constitute criminal child neglect) show a variety of unusual behaviors. As young animals, they explore little and play less. As adults, they are fearful and aggressive in response to threats and have a strong HPA response to separation from their peer group.

This phenomenon is especially severe in monkeys that carry a particular variant of the serotonin transporter gene, also called 5-HTT, which is also reported to increase vulnerability in people. The serotonin transporter removes serotonin from the synapse after its release by neurons, terminating the neurotransmitter's activity. Among primates, only people and rhesus monkeys show variation in this gene. Any given gene comes in different versions called *alleles*, which are part of normal genetic variation. Monkeys carrying the short allele of the serotonin transporter show higher HPA reactivity than animals with two copies of the long allele, but only when raised in a peer group. When raised by mothers, both types of monkeys have similar adult outcomes. Researchers conclude that variation in this gene contributes to behavioral flexibility in the population, which in turn allows rhesus monkeys—and people—to thrive in many different environments.

What determines individual stress responses in people? Recall from chapter 4 that specific combinations of genetic and environmental factors can lead to certain outcomes when neither factor alone has an effect. These interactions are challenging to tease out because so many combinations are possible.

One major contributor to our understanding has been a longitudinal study that followed over one thousand children born between April 1972 and March 1973 in Dunedin, New Zealand, from birth into adulthood. In this study, Avshalom Caspi, Terrie Moffitt, and their colleagues identified several psychiatric examples of gene-environment interactions, in which an experience increases the risk of a certain outcome only in people who carry a particular gene variant.

Children seem to develop their coping skills most effectively if they are exposed to a moderate amount of stress: high enough that they notice it, but low enough that they can handle it.

The best-known finding of the Dunedin study is that the risk of depression following childhood maltreatment or stressful life events is higher in people with the short allele of the serotonin transporter gene (and higher still in people with two copies of the short allele). People with two copies of the long allele are most resilient, with a low risk of depression regardless of their childhood experience. In contrast, people with two copies of the short allele have a risk of depression that increases in proportion to the childhood abuse they've suffered. Severe maltreatment doubles the risk of depression in this group.

The Dunedin study uncovered two other dangerous gene-environment combinations. Childhood maltreatment is more likely to lead to antisocial behavior in people who carry an allele that leads to low activity of a variant of the enzyme **monoamine oxidase**. This enzyme degrades serotonin, dopamine, and norepinephrine, all neurotransmitters involved in stress responding and mood regulation.

Also, heavy marijuana use during adolescence increases the risk of

PRACTICAL TIP: DANDELION AND ORCHID CHILDREN

If certain genes make children more vulnerable to damage under stressful conditions, why do they persist in the population? Some researchers think it's because children who are more sensitive to the environment may fare better in stable, supportive conditions—though conversely, difficult environments are harder on them. That is to say, we might replace the label *difficult children* with *orchid children*: they're sensitive, but if raised well, the results can be great.

As we already mentioned, most children are dandelions; they flourish in any reasonable circumstances. In contrast, orchid children with difficult temperaments (quick to anger or fear) benefit measurably from supportive parenting (see chapter 17). The combination of difficult temperament and harsh or unreliable parenting is a strong predictor of delinquency and psychiatric problems in many studies.

Under good rearing conditions, the outcomes for these children may be better than for dandelion children in many ways. For example, in one prospective study of high-stress families, high-reactive children were sick much more often than low-reactive children in the same families. In low-stress conditions, the advantage was reversed: high-reactive children were sick less often than low-reactive children. In short, outcomes for orchids were more variable, for good and for bad.

Similar results were found in an experimental study with rhesus monkeys. Highly reactive infants raised by especially nurturing mothers showed the most resilience in response to stress and ended up high in the colony's dominance hierarchy. High-reactive infants raised by average mothers had the worst outcomes. For infants born to average mothers, in contrast, being raised by nurturing or average mothers had little effect on adult behavior.

The orchid idea is consistent with several other findings. For example, inborn difficulties with reading could be associated with greater artistic capacity (chapter 25), and an interest in science might be associated with susceptibility to autism (chapter 27). Parents can take heart from the idea that the special care that's necessary in handling their difficult children may lead to good outcomes later on.

schizophrenia in people with a particular variant of the gene for *catechol-O-methyltransferase.* This enzyme breaks down the neurotransmitters dopamine, epinephrine, and norepinephrine. People who start using marijuana as adults do not have an increased risk of schizophrenia, regardless of genotype.

In general, gene-environment interactions are complex and thus difficult to demonstrate, so all these findings remain somewhat controversial. The interaction between the 5-HTT gene and early life stressors is the strongest result. It has been replicated in sixteen studies, though some others drew different conclusions. The marijuana-schizophrenia link is the weakest, resting on a single study so far.

The study of resilience in the face of adversity has contributed a great deal to our understanding of how experience affects child development. It's become increasingly clear that some environmental influences are individually variable, having different effects on different children. We know that life events can lead to different outcomes depending on which genes a child carries. There are likely to be even more complicated interactions that researchers cannot identify with current techniques, such as those involving multiple environmental conditions and multiple genes. Scientists may never unravel all these influences, but we can say for certain that neither genetic inheritance nor environmental conditions alone are destiny for any child.

Chapter 27

MIND-BLINDNESS: AUTISM

AGES: ONE YEAR TO FOUR YEARS

Most babies are interested in socializing from birth, but for children with autism, the social aspects of brain function are severely impaired. As we described in chapter 2, the first steps in development are programmed by genetic mechanisms and go awry only in cases of severe environmental flaws such as high doses of toxins. Sometimes, though, flaws in the genetic program can have a profound effect. Development is a complex business, depending on thousands of genes. When by chance children inherit unlucky combinations of alleles from one or both parents, neurodevelopmental problems such as autism can result.

At autism's core is a distinctive and profound disorder. One way to describe the problem is a deficit in *theory of mind*, the ability to imagine what other people know and what they are thinking or feeling (see chapter 19). Autistic people have difficulty recognizing when others are lying, being sarcastic, mocking them, or taking advantage of them. They have particular trouble in reading facial expressions, especially of emotion. In analogy to traits such as color-blindness, autistic children seem to have "mind-blindness."

Autism as originally defined in 1943 by its discoverer, Leo Kanner, is relatively rare, affecting one or two children per thousand born in the U.S. It is diagnosed based on three major problems: impaired social interactions, defective or absent communication, and repetitive or restricted behavior. Since then, other researchers have found that autism is not the same in all cases. Many children with autism also have other types of disrupted brain function such as mental retardation or epilepsy. Related to autism are less severe disorders such as

Asperger's syndrome, in which language and cognitive ability are largely intact. Collectively, these so-called autism spectrum disorders affect about one in 150 children, 75–80 percent of whom are boys. Siblings of these children are also at risk: they are almost ten times more likely to have an autism spectrum disorder than the general population (though the risk is still only one in twenty).

The recorded rate of autism spectrum disorders has increased dramatically in the last few decades. Much of this increase is due to changes in diagnosis. Autism was not recognized as a formal diagnostic category of the American Psychiatric Association until 1980, and a 1994 revision of the criteria made a larger fraction of children eligible for the diagnosis. In addition, more children are being screened, as many pediatricians now routinely administer questionnaires to identify autism. Some researchers believe that these factors can account for most of the apparent rise in the rate of autism.

Other possible contributors are increases in birth before full gestation, for

SPECULATION: ARE FERAL CHILDREN AUTISTIC?

Mythology and modern history contain many accounts of feral children. Typically these children are reported to lack the power of speech, usually grunting or howling instead. They reject clothing, reminiscent of perceptual difficulties in autistic persons, and are unable to socialize normally with others. Could it be that these children are not feral but autistic?

Such children are often said to be raised by animals, most often wolves but sometimes dogs, apes, bears, and even the occasional gazelle. Just as developmentally disabled children are mostly male, "wolf-reared children" are nearly always boys, whatever the country of the tale's origin. Romulus and Remus were said to be suckled by a she-wolf. Other such children in ancient Greek and Roman lore are usually male, though some cases are highly fanciful, such as a god being nursed by bees. A more modern example is Victor the Wolf Boy of Aveyron, who appeared to be about eleven years old when he was found naked and wandering in the French countryside in 1800.

Of course these children are never found in the actual custody of animals; their improbable rearing is reflected in their behavior. In the few cases where children can speak, they often describe running away from home and joining a group of animals. It seems more plausible that an abandoned, high-functioning autistic child could manage to survive long enough to be found.

instance, due to early induction of labor, and decreases in infant mortality due to improved care for premature and at-risk babies (see chapter 2). Unlike changes in diagnosis, such birth-related causes can lead to genuine increases in the number of children with neurodevelopmental problems. Another possible cause is increasing parental age over the last few decades, which may lead to more spontaneously occurring genetic changes in sperm and egg. The overall result is a boom in reported cases—and in public awareness.

Signs of the social and communication deficits in autism can be seen at one year of age or even earlier. At this age, infants with autism are less likely than nonautistic mentally retarded infants to respond to their own names, to look at people, or to use gestures to communicate. There are striking deficits in joint attention (shared attention with another person) to objects that people are holding, an early precursor to social behavior (see chapter 20), along with an inability to understand simple games such as peekaboo and sometimes tickling. These differences are relatively subtle, though, and few babies are diagnosed this young unless they have an autistic older sibling, a situation in which parents may be on the alert for unusual behavior. Autistic infants do form attachments to their parents (though these feelings may be expressed in unusual behaviors) and respond differently to strangers than to familiar people. In the second year of life, many autistic babies begin to have obvious developmental problems, such as language delay or repetitive behaviors.

Autism can be detected in a visit to the pediatrician or a developmental specialist, typically at twenty-four months. At that age, pediatricians often screen using a checklist for autism in toddlers. The Autism Diagnostic Observation Schedule survey can be given as early as twelve months, but there is some risk of making a premature interpretation. Some children simply take a little longer to develop but end up fine. By the third birthday, diagnosis is reliable.

It has been a challenge to researchers to understand how this strange combination of symptoms can arise. From a number of studies of brain structure at the levels of whole structures and individual cells, a few clues have emerged. There is still some ambiguity of interpreting results since typically brain abnormalities are observed after death, decades after the initial recognition of autism. It is unclear whether the abnormalities are present in early childhood, or whether they arise after years of the disorder's progress. Still, in combination with brain scans, the following patterns are found.

In scans of the brains of living autistic people, a common feature is a malformed or small amygdala. The number of neurons in this region is also decreased compared with nonautistic persons. The amygdala is necessary for the generation of emotional reactions (see chapter 18) and is activated when people are asked to evaluate emotion in other people's facial expressions, something autistic children have difficulty doing.

The structures most consistently affected in autistic brains are the cerebellum and other regions that are connected to it. In terms of numbers of neurons, reductions in cerebellar structures are seen in three fourths of autistic brains. Abnormalities are seen in whole-brain scans as well. When pediatric neurologists examined premature babies to look for a relationship between neurodevelopmental problems and the location of any brain injuries, they found that autism was specifically associated with damage to the cerebellum, more than to other brain regions.

These findings provide potentially useful clues to what makes autistic brains different—but also present a puzzle to researchers. Traditionally, the cerebellum was thought to be principally involved in processing sensory information to guide movement, a function that is disrupted when the cerebellum is damaged in adults. Autistic children are sometimes clumsy, but not in a debilitating way. So what's going on?

One possibility is that the cerebellum is essential for translating sensory events, such as the sight of a mother's smiling face, into a message with social import. Recall that your child's brain goes through phases of experience-expectant development (see chapter 10), in which the brain is ready to wire itself up, as long as it receives normal input. The brains of autistic children may have trouble translating everyday social experiences into a meaningful signal—thereby depriving themselves of a necessary experience early in life. If an abnormal cerebellum is involved in this derailed developmental process, it might do so through its connections to other brain regions, which include the anterior cingulate cortex, which is involved in face processing, and the prefrontal cortex, which is involved in complex planning and executive function. These brain regions also often show abnormalities in autistic people.

Abnormalities in the cerebellum may also cause deficits in perception. The cerebellum is important for distinguishing between touch from oneself and the touch of others. A prominent example is the phenomenon of tickling: you cannot tickle yourself because your brain generates signals large enough to feel the sensation only when it comes from another person. Activity is increased in the cerebellum during touch from others.

Evidence that autistic toddlers experience social sensory experiences in an unusual way comes from watching their reactions. They show difficulties with detecting natural biological motion such as walking, as well as with interpreting common social cues. When adults speak, typical toddlers look at the speaker's eyes, which convey social information. Autistic toddlers tend to look directly at the mouth, where the sound is coming from. In one study, toddlers saw a movie, with a soundtrack, of lights attached to various points on a model's head and body while he played a make-believe game with a teddy bear. Autistic toddlers spent equal amounts of time looking at the video played forward and the same video played upside-down and backward, so that the audio and visuals were mismatched. Typical and nonautistic disabled children focused mostly on the right-side-up movie.

Perceptual difficulties persist throughout development and into adulthood. Autistic people often show a high degree of sensitivity to routine sounds and even the sensation of their own clothing. Temple Grandin writes of her own experience as an autistic person: "Loud noises were also a problem, often feeling like a dentist's drill hitting a nerve. They actually caused pain. I was scared to death of balloons popping . . . Minor noises that most people can tune out drove me to distraction . . . My roommate's hairdryer sounded like a jet plane taking off." L. H. Willey, who has Asperger's syndrome, writes: "I found it impossible even to touch some objects. I hated stiff things, satiny things . . . Goose bumps and chills and a general sense of unease would follow."

At autism's core is a deficit in *theory of mind*, the ability to imagine what other people know and what they are thinking or feeling.

In identical twins, if one twin is autistic, the probability that the other is autistic is between 60 and 90 percent. Nonidentical twins, who share only half their genetic material with each other, have a much lower rate of diagnostic concordance. These facts suggest that autism's roots are genetic, and that they involve multiple genes. Researchers have been able to find warning signs as early as one to four months of age, indicating that developmental consequences of such a genetic inheritance can appear very early in life.

In recent years, genes that are associated with a higher likelihood of autism have begun to be found. Some of these genes are involved in brain development. Others encode proteins that are found at synapses, suggesting that they may affect the development or some other function of synaptic connections. Often one of these genes is present in some quantity other than the usual two (one from Mom and one from Dad), in a phenomenon called *copy number variation*. Although it is not clear exactly how these genes increase the risk of autism, in most cases autism is caused by combinations of genes. It may be that if neurodevelopmental processes encounter multiple difficulties, a sufficiently large perturbation will

PRACTICAL TIP: BEHAVIORAL THERAPY IS HELPFUL IF STARTED EARLY

Many therapies have been tried to treat the problems of autism, ranging from the mainstream to the speculative (see p. 239). The evidence that these treatments work is weak at best. The notable exception is intensive behavioral therapy, for which the most evidence for efficacy exists. Intensive behavioral therapy takes several forms and can help about half of autistic children.

In the 1970s and 1980s, Ivar Lovaas and his collaborators at the University of California, Los Angeles, developed a therapy for autistic children consisting of intensive one-on-one instruction, supervised training play sessions with typical (unimpaired) children, inclusion in regular classroom activities, and parent training for further at-home therapy. This intensive approach improves function in autistic children.

Behavioral therapy can begin as soon as autism is diagnosed, usually at age two or three. The UCLA model starts by getting children to respond to unambiguous instruction, starting with simple tasks. Correct responses are initially rewarded by foods and desirable sensory and perceptual objects. Later the children feel rewarded by praise, tickling, hugs, and kisses. Once they can answer questions, take turns, and engage in basic play, they are paired with typical children who give feedback during supervised play. Eventually the autistic child is introduced to classroom situations and group play.

Compared with other therapies or regular special education, children receiving behavioral therapy are more socially engaged and have better language, with improvements in IQ averaging twenty points. However, behavioral therapy is expensive and arduous. The cost is approximately $50,000 per year. Even this large cost may be less expensive than caring for an autistic person over his or her lifetime. Therapy requires at least thirty hours per week of direct attention by clinic staff or by parents working under staff supervision. Focus is required; in moderately impaired autistic children aged four to seven, combining behavioral therapy with other approaches has been reported to be far less effective, perhaps because efforts to give the effective therapy are diluted.

Given the developmental history of autism, it would seem logical to

start therapy as early as possible. One research team described Catherine, the one-year-old sister of an autistic child. Catherine obsessively closed open doors, had little language, and spent much of her time balancing long objects such as rulers vertically on her hand. Catherine underwent intensive treatment for three years, after which she entered regular kindergarten and tested above average in cognitive and language skills. At the end, Catherine blended in with typical children. Efforts are now under way to test the effectiveness of very early intervention.

Although this anecdotal case cannot prove whether Catherine would have otherwise become autistic, her successful outcome does suggest the potential benefit of identifying children at risk for autism before the age of two. Catherine's case may also demonstrate that a sufficiently mild dose of autism susceptibility genes may carry "orchid" advantages (see *Practical tip: Dandelion and orchid children*, p. 228) under suitable conditions, as evidenced by her eventual above-average performance.

---

cause the brain to veer away from a typical path and toward autism spectrum disorder. Dozens of susceptibility genes have been found so far.

One question is why the genetic factors that underlie autism would persist in the population. After all, aren't these the kinds of defects that disappear through evolution? This is generally the case, but an important exception occurs when combinations of genes cause a problem but confer some benefit individually. A well-known example is the gene for the oxygen transport protein hemoglobin. When a child inherits one copy that is altered in a particular way, he has increased resistance to malaria. When both copies are altered, sickle-cell anemia results.

Similarly, individual autism risk genes may have other functional effects. For example, autistic people tend to be very good with details, perhaps because of a lack of higher control from the frontal cortex. A small number of people in the population with an exceptional ability to focus on tasks could be a good thing for those people—and for society. In the words of Temple Grandin, "What would happen if the autism gene was eliminated from the gene pool? You would have a bunch of people standing around chatting and socializing and not getting anything done."

Another consequence of carrying autism susceptibility genes is an enhanced likelihood of interest in technical fields. Sam recently surveyed an entire entering freshman class at Princeton University. He found that among students expressing an interest in a technical field (science, engineering, or mathematics), one in twenty-five reported having a sibling with autism spectrum disorder. This rate was over three times as high as that found for aspiring humanities and social science majors, one in eighty-two. Similarly, in a previous study, physics and math majors at the University of Cambridge reported having relatives with autism spectrum disorder far more often than English and French majors did. These findings suggest the possibility that autism susceptibility genes might lead to a predisposition to a particular characteristic of thought, such as looking for systematic explanations for events in the world (or conversely, a predisposition not to think in terms of social explanations).

Autism is triggered largely by genetics.

Autism does not occur together in every pair of identical twins, suggesting that environmental causes may also contribute to the disorder. These environmental effects probably act during early infancy or even before birth. One example is the drug Depakote (valproic acid), which is given for epilepsy and psychiatric illness. It can increase the rate of autism when mothers take it during pregnancy. Another example is prenatal stress in the fifth, sixth, or ninth month of gestation, which is associated with a higher rate of autism (see *Practical tip: Less stress, fewer problems*, p. 16).

Despite reports to the contrary, it is a myth that autism can be triggered by vaccination. In the 1990s, specific blame was leveled at a particular vaccine, MMR, which is typically given at the age of twelve months. The original report was widely covered in the popular press. However, the paper was retracted and found to be fraudulent, events that have garnered less media attention.

A detailed investigation showed that the researcher Andrew Wakefield falsified the medical records of every child in the study, for instance, concealing the fact that children showed signs of autism before being vaccinated or were not autistic at all. Nearly all of the children were referred to Wakefield by a lawyer who was paying him to conduct the study to aid a lawsuit against vaccine manufacturers. Wakefield was stripped of his clinical and academic credentials, but continues to push a vaccine-autism connection in the U.S. with supporters such as the celebrity Jenny McCarthy. In several communities where the MMR vaccine has been withheld, the rate of autism has stayed the same, and in some cases the rate has increased. Since the developmental steps leading to autism are already well under way before the age of one, the main effect of withholding vaccination is to increase the risk of disease to your child and to his friends.

Even though the genetic causes of autism are starting to be understood, effective treatment is far off. Currently, autism has no silver bullet. One treatment that yields some positive outcomes is behavioral therapy. Unfortunately, this can be a hard road (see *Practical tip: Behavioral therapy is helpful if started early*, p. 236).

Some of parents' motivation to determine the causes of autism and seek treatment for their children may come from a feeling of guilt, even though they are not at fault in any way. Although these efforts are essential in driving forward research into causes and treatments, parents should be careful about embracing unproven and pricey new treatments, many of which have no benefit. Examples include chelation and nutritional therapy, facilitated communication, and hyperbaric oxygen treatment—the list goes on. These treatments are unlikely to be any better for a child than doing nothing, and in some cases they carry a significant risk of harm (see *Practical tip: Spotting untrustworthy treatments*, p. 244).

For autistic children, the strongest contribution that parents can make is to recognize the potential for problems early in life—and to act by the age of two, or earlier if possible. In most cases, babies simply mature on their own timetables, but for this disorder, intervention can make a critical difference. The challenges posed by autism spectrum disorder are considerable and will be with us for the foreseeable future. However, considering the possible benefits seen in their non-autistic relatives, some of the genes that make children autistic may also help others to thrive and contribute to society.

Chapter 28

OLD GENES MEET THE MODERN WORLD: ADHD

AGES: EIGHT YEARS TO EIGHTEEN YEARS

When Charlie Gross was a boy in the 1950s, he was deemed hyperactive. Although he was bright, his teachers bored him, so he was considered to be a troublesome child. He found other challenges, becoming an Eagle Scout and winning the Westinghouse Science Talent Search as a high school student. Years later, as a leading brain researcher, he was the first to discover single neurons in primate brains that respond to complex features, such as faces (see chapter 25).

If he had been born fifty years later, there is little doubt that he would have been offered treatment for ADHD. This disorder is estimated to occur in 5 percent of children worldwide. A looser definition gives an estimate as high as 17 percent, and in some school systems, up to 20 percent of boys receive drug treatment for this disorder. Is a phenomenon that is so widespread—yet so variable in estimated occurrence and treatment—a clearly defined medical disorder?

In many respects, the answer is yes, as demonstrated by treatability and the recent identification of genes associated with it. However, ADHD's history—including the recency of modern conditions that brought it to the forefront of public attention—has led to a fair amount of skepticism. Definitions of ADHD have changed several times since 1980 and are not always applied consistently. According to the American Psychiatric Association's *Diagnostic and Statistical Manual of Mental Disorders* (DSM-IV), evaluators ask whether a child "is often easily distracted by external stimuli," "often talks excessively," or "often blurts

out answers before questions have been completed." Most children (including Sam in his earlier years) meet at least some of these criteria. Indeed, your child may have some of these characteristics, which are common in children. However, diagnosis requires that these symptoms must exist to a degree that is considered maladaptive.

The strongest evidence that ADHD is a real disorder comes from genetics. The susceptibility to ADHD is inherited. In genetic linkage studies, ADHD has a heritability of 70–80 percent, on par with autism and greater than schizophrenia. There are dozens of identified ADHD susceptibility genes, many of which are involved in development and are also associated with autism and schizophrenia. Indeed, ADHD brains show changes in the growth trajectories of gray and white matter compared with normal development (see chapter 9).

At the same time, ADHD is also a product of social and cultural pressures. Your child's brain was originally optimized by natural selection to help him handle everyday problems, which did not include sitting in a classroom, much less resisting the attraction of a television or text message. The mismatch between evolution and civilization has not always been addressed by treatment. Long ago, children who made trouble often dropped out of school and sometimes society—think of Huckleberry Finn. Some of them ended up working on farms or drifting into crime. In most developed countries, we no longer let such children go down their own paths. Stimulant drugs and other therapies provide a means to treat such children—perhaps along with other children who do not belong in this category.

ADHD appears to be part of the natural range in attentiveness that results from generation-to-generation shuffling of the gene pool.

Attention-deficit hyperactivity disorder is misnamed; children with this condition have the capacity to pay attention, but they lack the ability to control where their attention goes. Many such children have problems with executive function, a suite of capacities that includes planning ahead, inhibiting undesirable responses, and holding information in working memory (see chapter 13). One consequence is that ADHD children are bad at estimating time intervals of up to a minute, missing wildly. A second area of deficit is an inability to forgo a small immediate reward in order to get a larger one that comes later. For this reason, they count future rewards less than other children do when they make decisions about what to do.

Teachers and parents may be motivated to look for ADHD by the availability of drug treatments for improving children's ability to focus. Most prominent among these is methylphenidate. This drug was first synthesized in 1944 by the chemist Leandro Panizzon. His wife, Rita, who had low blood pressure, used the drug to pep herself up before tennis games. In a romantic gesture, Leandro named the drug Ritaline after her—today, Ritalin. In addition to its alerting qualities, Ritalin also helps mental focus and began to be used for ADHD in the 1960s.

Ritalin's major biological action is as a **dopamine uptake blocker**; it prevents the neurotransmitter dopamine from being taken back up into neurons after it is released by synapses, thus prolonging its action on its receptors. Neurons in the ventral tegmental area and the substantia nigra release dopamine for a wide variety of functions: to regulate movement, to signal a rewarding event, and to control attention (see chapter 14). Dopamine uptake blockade is also the mechanism by which cocaine and amphetamine act. When cocaine, amphetamine, or Ritalin is present, dopamine hangs around longer in brain tissue and reaches higher concentrations—thus providing a stronger signal and better control over attention.

Only a few genes related to dopamine signaling have been linked to ADHD, and these genes have only a small influence on whether children develop problems. So despite the effectiveness of Ritalin, ADHD is not necessarily caused by a dysfunction of dopamine signaling. A more plausible explanation is that developmental steps impair the brain's ability to stay on task, with dopamine signaling as one mechanism that feeds into the circuitry. Differences in this circuitry have been probed by measuring both activity and size of key brain structures.

Distinctive patterns of brain activity are seen in ADHD children. Electroencephalography (EEG) can record electrical signals at the scalp that reflect the activation of neurons near the recording electrode, in approximate synchrony. At this broad level, brain activity oscillates at a variety of characteristic frequencies, with different frequencies becoming more prominent depending on the task at hand. For instance, the theta rhythm, which rises and falls between four and seven times per second, is active in idling brains—a signal that indicates that a person is spacing out. Higher frequencies include the alpha (eight to twelve times per second) and beta (twelve to thirty times per second) rhythms, which become more prominent in a variety of states including relaxation, inhibition of action, and alert concentration. All of these can be measured using EEG.

In children and adults with ADHD, alpha and beta rhythms are smaller in strength relative to theta rhythm than in typical children. The disparity between these rhythms occurs in ADHD children resting with their eyes open or closed, as well as when they engage in other activities such as a drawing or solving a

PRACTICAL TIP: SPOTTING UNTRUSTWORTHY TREATMENTS

We live in an age of celebrity spokespersons. For example, the model and comic actress Jenny McCarthy is an advocate for a popular—but thoroughly disproven—connection between vaccines and autism (see chapter 27). How should parents react to this onslaught of advice?

There is a long list of products sold using marketing claims that are poorly supported by science. These include brain scanning to diagnose and treat ADHD, balancing exercises for dyslexia, chelation therapy for autism, and nutritional supplements to aid brain function. Unfortunately, it's often difficult to distinguish between scientists with no financial interests and companies trying to manipulate their data to sell products.

Often a speculative treatment is based on some loosely related piece of evidence. For example, in many disorders such as autism, abnormalities are seen in the cerebellum, a structure that is traditionally associated with movement. Organizations such as the Dore Programme assert that movement exercises can alleviate all sorts of problems, including dyslexia, autism, and learning difficulties, but there is no credible peer-reviewed evidence for these sweeping claims.

When evaluating possible treatments for any problem, parents should ask this key question: is the argument for this treatment based on peer-reviewed literature or on inspirational stories? If it's based on stories alone, there is no reliable evidence for whether the treatment works. Other warning signs for quackery are the claim of a cure for a disorder whose causes are not understood, a single treatment that is claimed to be effective for multiple different disorders, and a failure to measure improvement objectively.

A few rules of thumb can help you identify which treatments are likely to be legitimate. Treatments that work for most people should be backed up by key phrases such as "peer-reviewed study," "controlled study," or "control group." When enough studies are done, meta-analyses can combine them into an even stronger form of evidence. If these elements are missing, all that is left are anecdotes—which do not guarantee that your child will obtain any benefit. Particularly if a Web site is dominated by individual testimonials or the authority of one person, watch out!

problem. It appears that the brain rhythms associated with idling are stronger relative to those associated with other mental states in ADHD children.

Based on these differences, it may be possible to improve function in ADHD kids' brains without resorting to drugs. Researchers devised exercises in which EEG signals are presented to the child as a form of *neurofeedback*. In a typical regimen, the child participates in a video-game-like exercise in which rewards are given for a desirable change, for instance, a decrease in the theta rhythm or an increase in the beta-to-theta ratio.

A meta-analysis of fifteen studies indicates that neurofeedback training reduces impulsivity and inattention considerably, with an effect size of 0.7 (see chapter 8 for a discussion of effect sizes); that's much larger than the improvements resulting from behavior modification alone and comparable to those seen using Ritalin. The meta-analysis included randomized trials (in which kids were assigned to treatment or no-treatment groups at random) and control groups who received similar amounts of training or therapist interaction as the test group, suggesting that the improvements came from neurofeedback treatment itself as opposed to other factors.

As might be expected from the EEG findings, differences are also seen in functional brain imaging studies. In this case, though, the differences are found when entire groups of ADHD children are averaged. Imaging methods are also far more expensive than EEG. Although some advertisements claim otherwise (see *Myth: The all-powerful brain scan*, p. 246), functional imaging is not reliable enough to be useful as a clinical or diagnostic tool.

On average, ADHD children show some subtle differences in brain structure from other children. Between the ages of six and nineteen, the brains of children with ADHD are 3 percent smaller on average than those of typically developing children. This difference is not uniformly distributed over all parts of the brain. The largest reductions are seen in white matter, which is made entirely of axons. White matter is reduced by 5 to 9 percent, suggesting that long-distance axons in ADHD children are narrower (and therefore slower) or reduced in number. There is also a slight thinning of the gray matter of the prefrontal and temporal cortex as well as the vermis, or central part, of the cerebellum.

One consistent finding in brain scans has been a reduction in the size of the *caudate nuclei*. These nuclei (left and right) form one component of the dorsal striatum of the basal ganglia, which communicate with many parts of the neocortex.

MYTH: THE ALL-POWERFUL BRAIN SCAN

A popular chain of clinics is operated by Daniel Amen, a cheerful celebrity doctor and bestselling author. At his Amen Clinics, expensive SPECT brain imaging scans are claimed to identify patterns of activity to design custom treatment for ADHD, anxiety, obesity, the prevention of Alzheimer's disease—and even marital problems. Amen's books are filled with vignettes in which the treatments are consistent with what most psychiatrists would do on hearing the patient's symptoms, without any brain imaging at all.

Scientists have offered to test the Amen Clinics' diagnostic tools under controlled conditions in which the evaluator makes a diagnosis without knowing the patient's problem in advance. Amen has declined this opportunity. Although neuroimaging may someday be useful in diagnosing brain ailments, no one yet has the ability to do it.

The basal ganglia are involved in directing attention and actions, for instance, in switching from one subject or task to another. One facet of switching is updating the importance of a particular stimulus or event. In addition, the basal ganglia select desired actions and reinforce the likelihood of action in future situations. Deficits in this ability could account for the difficulty that people with ADHD have in refraining from making an automatic or immediately appealing response—for example, looking in the direction of a distracting sound or attractive event.

The caudate receives powerful projections from the ventral tegmental area and the substantia nigra. The structural findings suggest that Ritalin may work by increasing the strength of dopamine signals coming into the caudate (and perhaps other brain areas as well). Think of the mechanism for staying on task as an unresponsive appliance switch, and Ritalin as a means for pushing the button a little harder.

Although ADHD is sometimes useful for identifying children who might need additional help, it is not a permanent designation. The signs of ADHD can change over time—mostly for the better. For example, though small children are generally not big on attention, in only the most extreme cases would it make sense to say that they have ADHD. Activity levels that may be quite typical in a four-year-old might be considered abnormal in a seven-year-old. In one estimate that is typical of the ADHD scientific literature, by age eighteen, ADHD symptoms have subsided in about 60 percent of boys who received the diagnosis earlier in life.

Most adults who once had childhood ADHD do not experience emotional or behavioral problems. In the long term, Ritalin does not improve academic outcomes—nor does it increase the risk for substance abuse, as you might fear for an amphetamine-like drug. In fact, the risk of later substance abuse may be lowered by Ritalin treatment. In general, ADHD kids are at higher risk for criminality and substance abuse, but the largest known predictor of this outcome is whether antisocial tendencies and conduct problems arise during adolescence (see chapter 9).

All in all, the boundary between ADHD and normal function is a blurry one that is determined by both biology and cultural expectations. To some extent, differences between ADHD and typical brains simply reflect a delay in development. In both groups of children, neocortical gray matter reaches a peak at or before the onset of puberty, but the peak occurs about three years later in ADHD children. Furthermore, the difference in the caudate essentially disappears by midadolescence.

These lags, as well as the resolution of behavioral problems in the majority of ADHD children, suggest that ADHD is at its core a matter of slightly slower brain maturation and that brains catch up by adulthood. In this respect, ADHD appears to be part of the natural range in attentiveness that results from generation-to-generation shuffling of the gene pool. Evolutionarily speaking, most of this range has generated functional people for most of the history of our species. Stimulants such as Ritalin should probably be reserved for those who are failing despite all other interventions. For other kids the right prescription may be, to paraphrase the old physician's advice: wait two years and call us in the morning.

Chapter 29

CATCH YOUR CHILD BEING GOOD: BEHAVIOR MODIFICATION

AGES: ONE YEAR TO TWELVE YEARS

When it comes to getting your kids to pick up their toys, we have good news and bad news. The good news is that children's behavior is strongly influenced by the positive or negative consequences that immediately follow from certain actions. If you can set appropriate expectations for behavior and get the consequences right (more on that later), your children will follow your household rules—most of the time, anyway.

The bad news . . . well, it's the same news. If whining or throwing tantrums gets your kids something they want, that's what they'll do. You may not think of nagging as a way of rewarding your child for misbehaving, but even yelling can actually encourage the behavior you're trying to stop, especially if that's the best way for your child to get your attention. Completely ignoring the problem behavior is usually the most effective way to get it to stop—if you can stick with it long enough.

It's common for parents to turn to yelling or spanking as their first response to problem behaviors, but a large body of research shows that this negative approach to behavior modification is not very effective in the long run. The effects of punishment are fleeting and tend not to generalize to other situations. Punishment also leads to fear and anxiety, which may cause emotional difficulties for

your kids down the line. Besides, you probably don't want to teach your kids that violence is an appropriate way to solve their conflicts with other people.

Many parents put extra pressure on themselves and their kids by believing that learning to obey family rules shapes children's eventual adult character. This belief may sound reasonable, but in fact, it's rarely true. No matter how you handle toilet training, your child is probably not going to be wetting the bed at twenty-five. As we discussed in chapter 17, parents do not sculpt children's personalities nearly as much as our culture leads us to believe. Understanding that your child's future is not at stake in routine parent-child conflicts, no matter how they turn out, should allow you to relax a bit.

The main effect of parents' rules and their consequences is to determine how your children behave while they're living with your family—and when they return home for holiday dinners as adults. Researchers find surprisingly little similarity between an adult's personality as evaluated by his friends or colleagues and his personality as evaluated by parents and siblings (see *Myth: Birth order influences personality*, p. 152).

If you're not building your child's character, why bother to enforce rules? There are several good reasons. Learning to abide by sensible restrictions does help children to develop self-control ability (see chapter 13). On the flip side, growing up in a chaotic household full of conflict is a common source of stress that can interfere with the development of resilience (see chapter 26). The most important reason, though, is simply that it's very difficult to build a good relationship with your children if you're trapped in a constant struggle over their behavior. Effective discipline allows everyone to put their energy into more important aspects of family life.

Explaining exactly what you'd like your child to do is the first step in behavior change—not the last, as many lecture-happy parents seem to believe.

A good foundation for a smoothly functioning household is warm parent-child relationships (see *Practical tip: Promoting conscience*, p. 177). Enjoying fun times together with your child is good for its own sake, of course, but it also helps to keep everyone on the same side, wanting what's best for each other. The easiest children to discipline are the ones who want to please their parents. Research shows that spouses who have fewer than five positive interactions for every negative interaction are at high risk of divorce. Despite the occasional temptation, parents and children cannot divorce one another, but a similar rule probably applies to distinguishing happy families from unhappy families. If you spend much of your time with your child nagging and correcting, it's worth giving some thought to how you can both get more enjoyment out of the relationship. That should improve both the quality of your home life and your child's behavior.

Broadly speaking, when parents talk about discipline, they want children either to do something or to stop doing something. One of the most effective methods of reducing the frequency of an unwanted behavior goes by the scary-sounding technical name of *extinction*, which is nothing more than ignoring the behavior. Your child whines; you act as if you didn't hear her say anything. Once you start that approach, though, you've got to hang tough until thc child stops whining. The very last thing you want to teach your child is that you'll give in after many hours of persistent whining—which is what a lot of kids end up learning, once their parents' resistance is worn down.

For that reason, it's easier to stop difficult behavior before it's become entrenched. You should be very skeptical of the phrase "just this once" when it pops into your head in moments of parenting stress. Frankly, whatever you're thinking of doing, you're unlikely to do it only once. So unless you're prepared to spend hundreds of hours in the car with your baby, don't use a car ride to put her to sleep. Similarly, putting your toddler to sleep by crawling into bed with him is setting yourself up for a lot of nights of human teddy bear duty.

Along the same lines, learning to anticipate and head off approaching problems before they become serious can save a lot of wear and tear on everyone. It's usually easier to intervene early by changing situations that lead your child to bad behavior—whether that means choosing the candy-free checkout line at the grocery store or suggesting that your child run around for a while before getting in the car for a long ride—than it is to deal with the resulting problems after they've occurred.

The practice of giving your child a time-out derives from studies of extinction in lab animals, where it is called *time-out from reinforcement*. As the name suggests, its purpose is to prevent children from getting any form of attention for bad behavior. Even negative attention is still attention. Like rewards, the time-out should immediately follow the behavior, or it will not be effective. Lecturing or touching your child during a time-out defeats its purpose and will probably act as an attentional reward. Brief time-outs of a minute or two are sufficient to change behavior. If at the end you briefly praise your child for cooperating with a time-out, she will be more likely to do so again the next time. If your child refuses to take a time-out when asked, it is time for her to learn the act of time-out itself. Use positive reinforcement on "dry run" time-outs while you are both calm (see *Practical tip: Getting to good*, p. 252).

In brain terms, extinction is not a form of forgetting but an additional form

PRACTICAL TIP: **GETTING TO GOOD**

It's more effective to reward your child for being good than to punish him for being bad. But how can you reward him if he won't do what you want him to do? There are two options, and you can use both of them together.

The first option is to reward him for being a little bit good. Imagine that you want him to pick up all his toys before dinnertime, but you can't reward him for that because he never does it. Instead of turning the interaction negative by nagging or giving up on the whole idea because it's too much hassle, set the bar lower at first. The first day, if he picks up a single toy, praise him immediately and enthusiastically, telling him exactly what part of his behavior made you so happy.

For young kids, there is no way to go too far with this strategy. Raise the roof with your praise—you should sound as pleased and excited as if he'd just bought you a new car.

"A little bit good" can also apply to complex actions such as tooth-brushing: for example, you can start by rewarding your child for simply holding the toothbrush. Over the next week or two, gradually set the bar higher, waiting until he's done a bit more before telling him how well he's doing, until he's brushing his teeth all by himself. Don't forget to continue to appreciate the good behavior after it's established.

The second option is to reward him for good behavior in a practice run. This approach is especially helpful for recurring situations that are difficult because either you or your child is too emotional for calm interaction. For example, if your morning routine is too stressful to allow time for behavior training, pick a moment when both of you are in a good mood and suggest that you play a game. If he successfully pretends to get his clothes on and come downstairs for breakfast, you'll give him a small treat. A few trial runs should pave the way for offering a similar reward for the real thing. For more details on using this approach correctly, see *The Kazdin Method for Parenting the Defiant Child*. In spite of the title, this book provides lots of helpful information for all parents on how to handle ordinary disciplinary challenges.

of learning. Laboratory studies show that extinction does not directly modify the synapses involved in the original behavior, but instead strengthens the frontal cortex's ability to suppress the existing activity of those synapses. As a result, the undesired behavior may suddenly pop up again at moments when frontal cortex function is weak—such as when your child is tired or has spent a long time focusing on something, like homework or chores. This outcome is expected and does not mean that the approach has failed, but it might mean that the kid needs a rest. As long as you don't reward the problem behavior, it will go away again.

Focusing your discipline on preventing negative behavior is ultimately a losing battle. To make changes stick, you also need to promote positive behavior, which is often simply the opposite of the behavior you want to remove. For instance, if your child is whining too much, it's not enough simply to ignore the whining. You have to also encourage positive behavior: when your child asks nicely, once, for what he wants, reward him. If he does that, even one time, jump on the opportunity to praise the behavior—and if possible, grant the request. Teaching your child a positive replacement behavior reduces the odds that the extinguished negative behavior will come back.

When an adult praises for small accomplishments, children over the age of six perceive it as a slight; they see the praise as reflecting the adult's low expectations.

Consistent, small rewards for small achievements work much better than large rewards for big goals, especially for younger children. After all, you wouldn't expect your child to learn to read if you paid him no attention until he'd finished his first book. Why set such a high expectation for behavioral self-control? Food and toys are often the first rewards that come to mind, but they are not the most effective. Your approval, expressed enthusiastically and accompanied by a pat on the shoulder or a high-five, should produce more behavior change than a cookie. Children also enjoy earning more control over their lives: the right to decide what's for dinner, stay up ten minutes later, or pick the destination for a family outing. These all make good rewards for positive behavior.

MYTH: **PRAISE BUILDS SELF-ESTEEM**

In the 1970s and 1980s, low self-esteem was held responsible for almost everything that could go wrong in a person's life, from fear of intimacy to child molestation to violence. As a result, government programs and private foundations worked hard to make children feel good about themselves. The idea was that because people with high self-esteem are happier, healthier, and more successful, encouraging the development of self-esteem would improve society.

Unfortunately, the research behind this belief suffered from many flaws. The most obvious was the problem of reverse causation: success makes people feel good about themselves, so of course successful people are confident. Another problem is that people with high self-esteem say a lot of positive things about themselves (that's pretty much the definition of high self-esteem), but many of those assertions are objectively incorrect. For example, people with high self-esteem rate their own intelligence as high but do not score better than average on IQ tests. In the end, interventions to improve self-esteem failed to improve academic achievement, job performance, or other objective measures of success.

The self-esteem movement had strong effects on parenting practice in the U.S.—but not necessarily good ones. Children do not benefit from routine empty praise, like the cries of "Good job!" that ring out over modern playgrounds. East Asian and South Asian parents (Sam's included) are known for strictness and are sparing in their praise, yet children from those cultures do not have particular self-esteem problems. Indeed, when an adult praises for small accomplishments, children over the age of six perceive it as a slight; they see the praise as reflecting the adult's low expectations.

Praise is most effective when it is specific and refers to something that your child can control. "You're so smart!" doesn't give your child any hint of what to do next time and may reduce perseverance (see chapter 22), while "Wow, you really worked hard on that math homework!" carries a clear message about the desired behavior. Parents who communicate high but achievable expectations, along with detailed guidance about how to get there, give their kids the tools to achieve real success in the world—which turns out to be the best route to self-esteem.

Explaining exactly what you'd like your child to do is the first step in behavior change—not the last, as many lecture-happy parents seem to believe. Prepare children for situations and let them know what is expected of them in advance. In the beginning, you should do whatever it takes to help your child succeed at earning the reward for good behavior, offering a cheerful reminder or two (but no more), standing in the room until the job is done, or even stepping in to help (without taking over the whole job). Such interventions provide scaffolding to support the behavior until it can stand on its own, but they should be temporary. If your child's behavior is improving, even slowly, then your efforts are working and you just need to stick with the program.

Parental inconsistency is a common cause of failure or slow progress. So is attempting to change too many behaviors at once. The best way forward is a systematic approach to working on one behavior at a time, while rewarding that behavior every time your child produces it. You won't make your best decisions in the heat of the moment; it's better to make a plan when you're calm, rather than trying to bribe your child on the fly with one-off promises and threats, which are notably ineffective.

Parents are human too, of course. Sometimes you're going to be tired or stressed yourself and fail to practice flawless disciplinary techniques—and that's okay. The occasional bout of yelling isn't going to do your kids any lasting damage. If reacting in the moment becomes your habit, though, you may be shortchanging your children and yourself. The next time you feel your temper getting the better of you, try stepping into another room and taking a deep breath. You might call it a time-out for grown-ups.

Chapter 30

A TOUGH ROAD TO TRAVEL: GROWING UP IN POVERTY

AGES: CONCEPTION TO EIGHTEEN YEARS

Growing up under conditions of deprivation can damage children's brains. This is an exception to the general principle we have expressed throughout this book that most children are resilient, and that variations in normal ("good enough") parenting do not appear to have a strong influence on how they turn out as adults. This chapter is about the other side of the coin: what happens to children whose developing brains match themselves to an environment that does not encourage them to express their full potential. After all, even dandelions can't grow without water.

Where your children grow up is one of the most critical factors in their development. When you move to a new house or apartment—or another country—you're determining not only your children's schools but also their neighborhood and the characteristics of the group from whom your children will select their friends. Children learn a lot from other kids and from the culture by which they are surrounded (see chapters 17 and 20). It's hard to raise children to reject the attitudes and assumptions of their peers, as parents have discovered everywhere from religious communities to inner-city neighborhoods. This is one of the many reasons that children start life at a disadvantage when they grow up in places with high unemployment, unsafe streets, and poor education.

Poverty itself isn't exactly the problem, unless children are actually starving, which is rare in developed countries. The risks instead come from conditions that are made more likely by poverty, in particular growing up in a chronic state

of fear and/or stress. Poverty is stressful due to a combination of economic insecurity (inadequate living conditions, frequent moves), disorganized households and harsh parenting (common side effects of parental stress or addiction), and social subordination (being treated as inferior because of social class and/or race). Heightened fear and anxiety can result from living in a high-crime neighborhood, food insecurity, and parental mistreatment (again more common when parents are stressed).

Inadequate parenting can and does occur in any segment of society, of course. Indeed, the middle class, because it is the biggest economic group in many countries, typically contains the largest number of chronically stressed or threatened children, as well as the largest number of children with behavioral problems. In addition, some especially resilient people who grow up in very difficult conditions become highly successful and happy adults. Even so, poor children grow up in environments that statistically increase their risk of a variety of disorders. Indeed, some of these "problems," such as chronic anxiety or early reproduction, may actually constitute adaptive responses to insecure living conditions (see chapter 26).

Socially and economically disadvantaged people are much more likely than middle-class people to suffer from medical, emotional, cognitive, and behavioral problems. *Socioeconomic status* (*SES*) is an umbrella term for the resources that people have available to them relative to others in their society. At minimum, it includes income, occupation (with associated prestige), and education, each of which can be broken down into more detailed measures. Across a variety of countries with different social systems, lower SES predicts substantially increased risk of a broad range of medical problems, including heart disease, respiratory disease, diabetes, and psychiatric conditions. As family SES decreases, children have increased risk of low birth weight, premature birth, infant mortality, injury, asthma, and various chronic conditions, including behavioral disorders. Community SES also influences child outcomes in studies that control for family SES.

People who are satisfied with their standard of living and feel financially secure are healthier, regardless of their actual income, occupation, and education, than people who are unsatisfied and anxious about the future.

Health and SES vary together across the full range of SES; the relationship is not merely a consequence of very poor health at the bottom of the scale. Overall, the lower people's SES, the earlier they are likely to die, with a difference of decades in some countries between the highest- and lowest-SES groups. The gradient is steepest at the bottom, though, with the biggest step between poor and working-class groups. These differences are large. In the U.S., adults with the lowest SES are about five times more likely to report having "poor" or "fair" health than the highest-SES adults.

SES is closely connected with health even in countries with equal access to health care and for diseases that medical care cannot prevent, such as juvenile diabetes and rheumatoid arthritis. So it is not primarily due to differences in medical care—though such differences can make the problem worse. Only part of this discrepancy (about one third, in one study of British government workers) can be explained by lifestyle differences, such as high rates of smoking and drinking, poor diet, and infrequent exercise among low-SES groups. Lung cancer is still more prevalent in low-SES than high-SES groups even when comparing people who smoke the same number of cigarettes, so there must be some problem beyond lifestyle choices.

The relationship between SES and health may be attributable to the effects of stress, which can damage the brain and the rest of the body (see chapter 26). In many species, life at the bottom of the dominance hierarchy involves chronic stress and a poorly functioning biological stress response system. You could imagine that animals with poor stress responses are just more likely to become subordinate, but researchers found that social subordination occurs first and causes poor stress responsiveness, and not the other way around.

It can be most stressful to be a high-ranking animal in some species or under special circumstances, for instance, when dominance can be maintained only by fighting a lot. But in people, it's usually the low-ranking members of society who experience the most stress. Social status is so important to people that reducing the power or status of middle-SES adults in an experimental situation decreases their ability to concentrate, ignore distractors, and inhibit inappropriate behavior. We speculate that chronically low social status may have a similar effect on low-SES children. In one study, by age ten, children in Montreal already showed a sharp relationship between SES and cortisol, with blood levels twice as high in the lowest-SES children as in the highest-SES children.

How we interpret the circumstances of our lives also has a strong effect on our stress responses (see chapter 26)—often stronger than the effects of our actual economic circumstances. Low-SES people not only experience more chronic stresses and negative life events, but also experience ambiguous events as being more stressful, compared to higher-SES people. When people are asked to give their own position in society on a drawing of a ten-rung ladder, their ranking is a stronger predictor of health than their actual SES. People who are satisfied with their standard of living and feel financially secure are healthier, regardless of their actual income, occupation, and education, than people who are unsatisfied and anxious about the future. Along the same lines, countries, states, or cities with greater income inequality have steeper gradients of SES versus health. This may be because income inequality interferes with the feeling of community, which provides many types of social support to counteract stress. Increased crime also correlates with income inequality—again, better than with absolute poverty. So the existence of strong inequality in society may be a major driver of stress.

Which parts of children's brains are damaged by deprivation? We know from animal studies that chronic stress can cause structural changes in the hippocampus and amygdala (see chapter 26). In people, low subjective SES and other sources of chronic stress are linked to reduced hippocampal volume. Long-term memory, which depends on hippocampal function, is impaired in low-SES populations. In experimental animals, chronic stress can cause neurons to die, prevent new neurons from being born or surviving, and cause dendrites to become less complex (a change that is reversible) in the hippocampus. Scores on a variety of language tests also vary strongly with SES, perhaps due to the less complex language environment provided by low-SES parents (see chapter 6).

In people, the perception of low SES is associated with stronger activity in the amygdala in response to threats. That's understandable; if you believe that you are low on the totem pole, it's natural to feel vulnerable and therefore respond strongly to danger. Indeed such increased vigilance may reflect a sensible reaction to real dangers in the environment. The amygdala is important for rapid processing of events that induce fear and other emotions (see chapter 18), and it is extensively interconnected with the stress response system.

Across the life span, from infants to adults, low SES predicts decreased executive function, perhaps because the environment offers fewer opportunities to strengthen these abilities through practice. The medial prefrontal cortex (including the anterior cingulate and orbitofrontal regions) is an important inhibitor of the stress system. In experimental animals and people, chronic stress reduces the size of the prefrontal cortex. This brain region is involved in working memory and planning and organizing behavior (aspects of executive function), and it is also necessary for learned suppression of fearful reactions to situations that are no longer dangerous. People who perceive themselves as having low SES have reduced volume in one part of the anterior cingulate cortex. One promising intervention for low-income preschool children, Tools of the Mind, focuses on promoting behaviors that depend on the prefrontal cortex (see *Practical tip: Imaginary friends, real skills*, p. 117).

The causes and possible solutions to the SES-health gradient are hotly debated, within the scientific community as well as in society. The key problem for research is that people aren't randomly assigned to be poor, so we can't draw conclusions about causality by comparing the characteristics of low-SES and high-SES people (see *Did you know? Epidemiology is hard to interpret*, p. 262).

Do people develop problems because they're disadvantaged? Or do they become (or remain) disadvantaged due to poor health or other problems? There is evidence in favor of both positions. The health of adopted children is best predicted by their adopted parents' income, not their biological parents' income, suggesting that family income can influence health independently of genetics. Along the same lines, childhood SES predicts adult health, as we discuss below. On the other hand, the adult income and (particularly) education of adopted children does depend partly on their biological parents' characteristics.

The existence of strong inequality in society may be a major driver of stress.

It's important to remember that these two classes of explanations aren't mutually exclusive. Indeed, the most likely relationship between poverty and achievement is a vicious cycle, in which starting life with few resources leads children to develop a variety of problems, which then make their life situation worse, reducing their resources (and their children's resources) still further.

Some of the relationship between SES and cognitive achievement may be attributable to exposure to environmental hazards, more common in poor neighborhoods, that can cause substantial, lasting impairment in brain function. Children exposed to lead before or during elementary school age have lower IQs and impulse control, as well as higher aggression and delinquency, compared with children of the same SES. All these problems persist through adulthood. Mercury exposure also reduces IQ, along with attention, memory, and language development.

Children who live in noisy environments, such as near airports or highways, are delayed in learning to read compared with other children of the same SES. Chronic noise exposure also causes deficits in attention and long-term memory, perhaps because it is known to increase stress hormone levels. Crowded or chaotic environments (at home or at school) impair cognitive development and academic

DID YOU KNOW? **EPIDEMIOLOGY IS HARD TO INTERPRET**

The tools of epidemiology, appropriately, are best suited to the study of epidemics, which are caused by a single factor (a germ). The same tools are increasingly used to study conditions like heart disease, which have far more complex causes. Epidemiological studies of this kind are far more difficult to interpret and should be approached with a skeptical eye.

In a typical epidemiology study, scientists collect data on a large group of people for years and then attempt to correlate risk factors, such as excessive drinking, with health outcomes, such as deaths due to injury. Studies of this sort have serious limitations, which are rarely taken into account in your local newspaper or when health agencies make lifestyle recommendations based on their findings.

It is almost impossible to draw reliable conclusions about cause and effect from correlation data. One pitfall is reverse correlation. For instance, obesity is correlated with poverty. Does poverty lead to poor diet and lack of exercise, which then cause obesity, as is commonly assumed? Or might obesity cause poverty due to wage discrimination against fat people? Another pitfall is that an additional (unstudied) factor might cause both parts of the correlation. Harsh parenting is correlated with later antisocial behavior. Does that mean harsh parenting causes antisocial behavior? Or could it be that some parents pass along a genetic tendency to antisocial behavior to their children, who then are likely to misbehave, evoking harsh parenting, even from adoptive parents? We did not invent these two examples. In both cases, there is good evidence for the second interpretation, at least as a partial explanation of the observed correlations (see p. 151 for more information on the effects of harsh parenting).

Making interpretation even more difficult, risk factors tend to travel in packs. Postmenopausal women taking hormone replacement therapy have fewer heart attacks than other women, but they are also less likely to die from homicide or accidents—effects that are unlikely to be caused by hormones. The explanation is that women who take hormone replacement therapy typically have a variety of healthy characteristics: compared to other women, they pay more attention to their health, exercise more, and

are richer, more educated, and thinner. When the risk factors are correlated with each other, it becomes very difficult to sort out causes from accidental "bystander qualities," even if the observed correlations are strong.

Epidemiology can be very useful. The link between cigarette smoking and lung cancer was established through this technique because the correlation is large (heavy smokers have twenty or thirty times more risk than nonsmokers) and the rate of lung cancer in nonsmokers is low. Many side effects of approved drugs have also been identified by epidemiology. But most lifestyle effects are small to moderate, and most of the common diseases in developed countries are influenced by multiple factors. Under those conditions, epidemiology can only generate hypotheses that must be tested by other means. Such studies should be interpreted with care and caution.

---

performance and increase psychological distress in both parents and children, again independent of SES. These environmental conditions are all common in the lives of low-SES children and often occur together.

Growing up in a low-SES family predicts poor health even for children whose SES improves in adulthood. For example, in a group of nuns who had been living together since early adulthood, disease risk and longevity still varied depending on their education (whether or not they had gone to college). For more than fifty years, the nuns had shared meals, housing conditions, and a very similar lifestyle, but the traces of their early experiences were still substantial, with educated sisters living an average of 3.28 years longer than less educated sisters. In general, people whose SES improves later in life gain less advantage from the change than people whose SES improves in childhood.

Children whose families move out of poverty improve in some areas but not others. One study followed 1,420 poor children in North Carolina from 1993 (at ages nine to thirteen) through 2000. American Indian families were more than twice as likely as non-Indian families to be below the poverty line when the study began. In 1996, a casino opened and began to distribute some of its profits to every person on the reservation. Children whose families moved above the poverty line showed a 40 percent decrease in antisocial behaviors during the study, while children whose families remained poor showed no change in antisocial behaviors.

In contrast, moving out of poverty had no effect on symptoms of depression and anxiety, though children who had never been poor had fewer symptoms than always-poor or ex-poor children.

If indeed poverty leads to a vicious cycle like the one we've described, it should be easiest to break that cycle in young children, before they fall too far behind their peers. Intensive preschool enrichment programs can have positive effects that last into adulthood, substantially increasing the odds of a poor child graduating from high school, finishing college, getting a skilled job, and owning a home. These programs can also reduce the likelihood that a child will need special education or repeat a year of school.

Mostly these effects do not depend on increasing children's IQs. Instead the positive outcomes seem to stem from improvements in social competence, including perseverance and motivation (see chapter 13) and emotional well-being. The programs that produce these results tend to be extensive, long-lasting interventions, which require a considerable commitment from both families and funding agencies. These programs are often still cost-effective for society in the long run if they reduce the likelihood that children will need special education or repeat a year of school or that they will receive welfare payments as adults.

Intervention is difficult for exactly the same reason that it is important: because it requires interrupting the developing brain's strong tendency to match itself to the local environment. As we've discussed throughout this book, evolution has made that matching process resilient and hard to disrupt. If a child's environment is toxic, though, it can do more harm than good. Fortunately, the reward for intervening is also large—turning that child into an adult who can function successfully in a safe and productive world, like the one we all want for our children.

ACKNOWLEDGMENTS

We could fill an elementary school auditorium with the people who made it possible for us to put together this book on children's brains. Before and during its writing, so many people generously shared their friendship, experiences, expertise, and time. We are grateful to them all.

At home, Sandra thanks her husband, Ken, for his talented care and feeding of authors and for his enthusiastic contributions to adventures large and small. She would also like to thank her parents for their unwavering dedication to raising the individual child that she was, rather than trying to push her into pursuing their own aims.

Sam thanks his parents, Mary and Chia-lin (Charlie), for a lifetime of love, dedication, nurturing, and teaching. His wife, Becca, has been a partner in so many ways: our life and adventures, critical reading of every page of this book, and raising our irrepressible daughter, Vita, a source of delight and lessons for both of us. Becca and Vita were also good-humored about Sam jetting off to Sandra and Ken's California eyrie for locavore cookery, long uphill and downhill conversation-filled walks, and the occasional bout of writing. Finally, the Princeton community of parents and colleagues was a great source of friendship, feedback, and support.

Lisa Haney and Patrick Lane again provided splendid illustrations, as they did for our previous book, *Welcome to Your Brain.* We are also grateful to Roger Tsien and Gordon Burghardt for permission to reprint photographs of chemistry apparatus and a playing turtle, and to Ken Britten for the image of what babies see.

For comments, conversation, anecdotes, and advice on various chapters, we are grateful to Ralph Adolphs, Robert Ammerman, Connie Ban, Daphne Bavelier, Dorothy Bishop, Gillian Blake, Paul Bloom, Ken Britten, Jeanne Brooks-Gunn, Silvia Bunge, Gordon Burghardt, BJ Casey, Anne Churchland, Karla Cook, Ricardo Dolmetsch, Chunyu (Ann) Duan, Barbara Edwards, Nancy Eskridge, Anne Fernald, Shari Gelber, Alan Gelperin, Anirvan Ghosh, Adele Goldberg, Alison Gopnik, Liz Gould, Charles Gross, Art Kramer, Eric London,

Bert Mandelbaum, Kim McAllister, Sara Mednick, Rebecca Moss, Rita Moss, Elissa Newport, Yuval Nir, Kathleen Nolan, Dan Notterman, Danielle Otis, Liz Phelps, Jessica Phillips-Silver, Emily Pronin, Robert Sapolsky, Steven Schultz, John Spiro, Lawrence Steinberg, Giulio Tononi, Marty Usrey, Anthony Wagner, and Jeffrey Wickens. By helping us with technical, medical, and child development facts and tone, they made us look better than we would have on our own. Needless to say, any remaining errors are our responsibility.

Our agent, Jim Levine, provided encouragement and reality checks as needed. He also helped us get in touch with Ellen Galinsky, who so kindly wrote a foreword that captures the spirit of the book. Beth Fisher didn't let a worldwide recession get in the way of connecting us with publishers in other countries. Indeed, everyone we worked with at the Levine Greenberg Literary Agency was committed and enthusiastic at every turn.

Our editor, Ben Adams, believed in this book from its conception and never wavered in his support, all the way through delivery. Thanks to the entire Bloomsbury team for their help and suggestions, especially to managing editor Mike O'Connor for shepherding the manuscript through production calmly and carefully.

Finally, we are grateful to all the parents who unwittingly helped by asking us questions about their children's brains. Your curiosity made this book worth writing. We hope you enjoyed it.

GLOSSARY

acetylcholine: A neurotransmitter whose functions include activation of muscles, and which is released from neurons in the parasympathetic nervous system, as well as within the brain.

action potential: A spikelike change in the voltage across the membrane of a neuron, lasting approximately one thousandth of a second and able to travel down the axon to its ends, where it triggers the release of neurotransmitters.

amblyopia: A disorder in which one or both eyes loses the ability to see details; sometimes referred to (along with another disorder, strabismus) as *lazy eye*.

amygdala: An almond-shaped structure under the rostral pole of the temporal lobe that is involved in basic positive and negative emotional responses, including fear.

anterior cingulate: The frontal part of the cingulate cortex, which surrounds the corpus callosum in a collarlike shape.

attachment: A strong and persistent desire to be close to a familiar caregiver, especially when the child is stressed or upset.

axon: A long, thin structure that emerges from a neuron, and which is specialized for the long-distance transmission of information by transmitting action potentials along its length toward its ends, where synapses reside.

basal ganglia: As part of the forebrain, a group of nuclei (clusters of neurons) located underneath the neocortex and involved in directing choices, attention, and rewards. The name is an exception to the principle that ganglia are defined as being found outside the brain.

brainstem: An evolutionarily old part of the brain that sits between the spinal cord and forebrain and controls basic functions that usually do not reach conscious awareness, such as breathing.

Broca's area: A part of the left hemisphere of the neocortex, discovered by Pierre Paul Broca, that is essential in the production and comprehension of language.

cerebellum: A brain component occupying about one seventh of the brain in most mammals, and which integrates sensory information to drive perceptions, movement, and higher functions.

cerebral cortex: See **neocortex**.

circadian rhythm: A cycle of brain and body activities that takes approximately one day, and which can proceed without day-night light cues.

cognitive: Relating to higher brain functions such as thinking, regulation of emotional responses, and declarative learning and memory.

congenital: Inherited by genetic mechanisms.

corpus callosum: The principal route for communication between the hemispheres of the neocortex; composed entirely of axons.

corticotropin-releasing hormone (CRH): A peptide released by the hypothalamus in response to stress, and which in turn activates the pituitary gland.

cortisol: A steroid hormone secreted by the adrenal gland. Cortisol is the principal human stress hormone or glucocorticoid.

dandelion child: A child who thrives in a wide variety of environments. Based on Swedish folk wisdom (*maskrosbarn*).

dendrite: A treelike structure, extending from the cell body of a neuron, that receives communicating synaptic inputs from other neurons.

dopamine: A neurotransmitter that regulates reward, attention, and action, and which is secreted by neurons in the substantia nigra and the nucleus accumbens, two small structures in the brain's core.

dopamine uptake blocker: A chemical, such as methylphenidate (Ritalin), cocaine, or methamphetamine, that prevents dopamine from being taken back up into neurons after it has been released, thereby prolonging dopamine's action.

dorsal: In the brain, the direction toward the top of the head; in the spinal cord, toward the back. The opposite of ventral.

effect size: The difference between groups divided by the variability of one or both groups. For Cohen's d' as described in chapter 8, d' = 0.2–0.3 is considered small, d' = 0.5 is considered moderate, and d' = 0.8 or larger is considered large. A moderate effect size would be likely to be noticeable in an individual in everyday life.

epigenetic: Having to do with inherited change that comes from mechanisms other than the DNA sequence itself. Here in particular, a long-lasting chemical modification to DNA that affects gene expression.

epinephrine (also known as *adrenaline*): A chemical signal from the sympathetic nervous system that activates fight and flight responses, and which is released from the adrenal glands.

executive function: A suite of related abilities for self-control that includes planning ahead, inhibiting undesirable responses, and holding information in working memory.

fMRI (functional magnetic resonance imaging): A noninvasive imaging method that uses the properties of oxygenated hemoglobin in blood to visualize where blood flow has increased in response to neural activity.

frontal lobe: A major part of the neocortex, located toward the front and containing a variety of smaller regions.

fusiform face area: A part of the visual system that is active in recognition of faces and other familiar objects.

gene: A sequence of DNA that encodes the sequence of a protein, as well as containing markers that specify under what conditions the protein will be made.

gestational age: The number of weeks since a pregnant woman's last menstrual period, approximately two weeks more than the number of weeks since conception.

glial cell: A cell of the nervous system that is not a neuron and that supports brain function. Plural, **glia**.

glucocorticoid: A class of steroid hormones involved in suppressing the immune response and secreted by the adrenal glands.

glutamate: The most widespread neurotransmitter of the brain, used by a neuron to increase the likelihood of other neurons firing. Also an amino acid, one of twenty used to make proteins and peptides.

gonadotropin-releasing hormone: A peptide molecule, secreted by neurons of the hypothalamus, that activates secretion of gonadotropin by the pituitary gland.

gray matter: The type of brain tissue in which neurons, dendrites, and synapses are found; like white matter, it can contain axons, blood vessels, and glia.

gyrus: A single fold of the neocortex, consisting of a part of the gray matter sheet folded on itself and containing a bit of white matter in its core. Plural, **gyri**.

hippocampus: A brain region central to learning, memory, spatial navigation, and regulation of emotional response.

hypothalamic-pituitary-adrenocortical (HPA) system: A complex set of interacting systems—including the hypothalamus, pituitary gland, and adrenal glands—that regulate stress responses.

hypothalamus: A brain region located beneath the thalamus with a central role in controlling many core functions, including emotional responses, stress, hunger, thirst, and sexual behavior.

insula (also known as *insular cortex*)**:** A portion of the neocortex deeply buried inside a sulcus between the temporal and frontal lobe; important in the processing of emotions and in the body's current state.

locus coeruleus: A small brain region that secretes norepinephrine and sends axons throughout the brain.

longitudinal study: A study in which the same people are followed over time, so that individual changes in function can be measured. As opposed to a cross-sectional study, in which people in different groups are compared with one another.

medial: The direction toward the body axis midline. The opposite of lateral.

melanopsin: A pigmented protein found in certain cells of the retina and involved in converting light into signals to the brain to drive the circadian rhythm.

meta-analysis: A statistical technique in which multiple studies are pooled to increase confidence in the overall conclusion and to detect biases in individual studies.

monoamine oxidase: An enzyme with multiple functions, including the breakdown of the neurotransmitters dopamine, serotonin, melatonin, epinephrine, and norepinephrine.

motherese: The distinctive speech instinctively used to speak to infants.

motor: Relating to movement.

mutation: An error in the copying of a gene.

myelin: A fatty sheath, generated by some glial cells, that wraps around axons to provide electrical insulation, thereby speeding signals.

neocortex (also known as *cerebral cortex*): The largest part of the human brain, occupying the great majority of the forebrain and three fourths of total brain volume.

nerve: A bundle of axons that lead to the brain, spinal cord, a muscle, or another organ or tissue.

neurodegeneration: A general term referring to progressive loss of structure and function of neurons.

neuron: Cells of the brain that process information and send it over long distances, including out to the body.

neuropeptide: A peptide used as a neurotransmitter.

neurotransmitter: A chemical or peptide used by neurons for signaling.

neurotrophins: A family of proteins that induces the survival, development, and function of neurons, for instance, by fostering the growth of dendrites.

norepinephrine (also known as *noradrenaline*): A neurotransmitter secreted by neurons of the locus coeruleus and used to send sudden signals to the rest of the brain concerning alerting to important events.

nucleus: (1) An organized cluster of neurons with clear boundaries inside the brain; such a cluster outside the brain is called a *ganglion*. (2) The center of a cell, where DNA is found.

occipital lobe: A major part of the neocortex, located to the back.

olfactory: Having to do with the sense of smell.

optic chiasm: A location underneath the brain and behind the eyes, where the two optic nerves meet and partially cross.

orbitofrontal cortex: A frontal part of the brain surrounding the eye socket (orbit).

parietal lobe: A major part of the neocortex, located slightly back from the top of the head on both sides.

peptide: A short chainlike molecule containing two to fifty amino acids; often used as a signaling molecule in the brain and body. Proteins are composed of longer chains (and sometimes lengths in this range as well).

plasticity: The capacity of neural tissue to change; synaptic plasticity is a change in the properties of synapses, such as the strength of their connection.

pons: A part of the brainstem found at a similar level as the cerebellum; bracketed by the midbrain above and the medulla below.

prefrontal cortex: The forwardmost part of the frontal lobe of the neocortex.

premotor cortex: A part of the neocortex in the frontal lobe, near the top of the head, just forward of the motor cortex; both are structures for planning and carrying out movement.

protein: A category of molecules found in all living organisms; a chain of amino acids strung together with a specific sequence encoded by a corresponding sequence of DNA. Proteins act in many roles, including as receptors, enzymes, and other vital cell components.

receptor: A protein that binds to other molecules, such as neurotransmitters, hormones, or other signals.

retina: A thin sheet of neural tissue found at the back of the eye.

rostral: Toward the front of the brain or spinal cord along an imaginary axis running from the forehead, bending at the base of the skull, and toward the tailbone. Its opposite is caudal.

sensitive period: A time in development when experience has a particularly strong or long-lasting effect on the construction of a particular aspect of brain circuitry and the behavior that it controls.

serotonin: A neurotransmitter secreted by neurons of the raphé nuclei of the brainstem and involved in mood, movement, and sleep.

somatosensory: Having to do with body surface sensation.

standard deviation: A statistical measure of the amount of variability in a population. For many common measurements, about two thirds of the measurements are within one standard deviation of the average, and 95 percent are within two standard deviations. For instance, if the average height of twenty-four-month-olds is thirty-four inches with a standard deviation of two inches, then about two thirds of these children are between thirty-two and thirty-six inches in height.

striatum: A subcortical region that receives input from the neocortex and provides input to the basal ganglia.

subcortical: A general term referring to most brain structures other than the neocortex.

substantia nigra: A component of the basal ganglia, containing neurons that synthesize dopamine.

sulcus: A groove in the neocortical surface between lobes or gyri. See **gyrus**.

superior colliculus: A brainstem region that is a major target for visual information; the superior colliculus is called the *optic tectum* in nonmammalian vertebrates.

suprachiasmatic nucleus: A nucleus located above the optic chiasm, and the master clock driving the circadian rhythm.

synapse: A junction between neurons where communication occurs, most often by the release of neurotransmitter from the axon of one neuron onto receptors in the dendrite of another neuron.

temporal lobe: A major part of the neocortex, located to the sides near the temples.

testosterone: A steroid sex hormone made in testes and ovaries; found in larger quantities in males than in females.

thalamus: A football-shaped structure at the brain's core, found under the neocortex and containing most pathways to the neocortex.

theory of mind: The understanding that other individuals have different knowledge and thoughts than you.

ventral: In the brain, the direction toward the bottom of the head; in the spinal cord, toward the chest. The opposite of dorsal.

ventral tegmental area: A group of midbrain neurons that secrete dopamine and that send axons throughout many regions of the brain; thought to serve functions relating to reward, motivation, and cognitive function. Near the substantia nigra, where dopaminergic neurons are also found.

white matter: A type of brain tissue composed entirely of axons, blood vessels, and glia, and whose myelin confers a white appearance.

NOTES

CHAPTER 1 The Five Hidden Talents of Your Baby's Brain

cats versus dogs: Quinn 2002
syllable boundaries: Saffran, Aslin, and Newport 1996
mobile and ribbon: Rovee-Collier and Barr 2001
properties of objects and agents: Spelke and Kinzler 2007
surprised the object was not solid: Baillargeon, Spelke, and Wasserman 1985
object permanence in three-and-a-half-month-olds: Baillargeon and Wang 2002
Freud discredited: Webster 1995
hands versus sticks: Woodward 1998
circle chasing circle: Gergely et al. 1995
male and female faces: Quinn 2002
early preference for faces: Johnson et al. 1991; Mondloch et al. 1999
early preference for voices: Fernald 1992
infant attention: Colombo 2001
adults influence baby's attention: Hood, Willen, and Driver, 1998; Reid and Striano 2005
for more on psychology in babies: Bloom 2004; Gopnik 2009

CHAPTER 2 In the Beginning: Prenatal Development

prenatal development: Sanes, Reh, and Harris 2005
autism and hurricanes: Kinney et al. 2008
ice storm and IQ: LaPlante et al. 2008
stress and schizophrenia: Khashan et al. 2008
prenatal drug exposure: Thompson, Levitt, and Stanwood 2009
fish and pregnancy: Hibbeln et al. 2007; Oken et al. 2008; Jones et al. 2009
prescribed drugs and developmental problems: Witter et al. 2009; Gentile 2011
neurodevelopmental disabilities, Norwegian study: Moster, Lie, and Markestad 2008
increase in premature birth: Goldenberg et al. 2008
increased survival contributes to preterm birth statistics: Ananth et al. 2005
thirty-four to thirty-seven-week births, elective induction of labor: Fuchs and Wapner 2006; Engle and Kominiarek 2008

CHAPTER 3 Baby, You Were Born to Learn

characteristics of breast-feeding mothers: Der, Batty, and Deary 2006
higher-quality studies less likely to find an effect: Jain, Concato, and Leventhal 2002

sibling study of breast-feeding and intelligence: Der, Batty, and Deary 2006
another study of sibling pairs: Evenhouse and Reilly, 2005; the authors claim to have found an effect of breast-feeding on cognitive function, but they use an unusually lenient measure of statistical significance ($p < 0.1$). The effect is not significant at $p < 0.05$, the typical scientific standard.
mothers randomly assigned to breast-feeding support program: Kramer et al. 2008
cross-cultural motor development: Adolph, Karasik, and Tamis-LeMonda 2009
practice accelerates motor development: Adolph, Karasik, and Tamis-LeMonda 2009
American nouns versus Korean verbs: Gopnik and Choi 1990
imitation using head or hands: Gergely, Bekkering, and Király 2002
infant brain development: Tau and Peterson 2010
head-eye coordination: Goodkin 1980
infants learn about objects: Gopnik, Meltzoff, and Kuhl 1999

CHAPTER 4 **Beyond Nature Versus Nurture**

epigenetic modifications: Zhang and Meaney 2010
epigenetic modifications in twins: Wong et al. 2010
epigenetic modifications passed along to offspring: Jablonka and Raz 2009
enrichment and inherited learning ability in mice: Arai et al. 2009
lactose tolerance and milk drinking: Holden and Mace 1997
culture drives evolution: Laland, Odling-Smee, and Myles 2010
heritability of IQ in different environments: Turkheimer et al. 2003
gene-environment interactions: Maccoby 2000; Rutter 2007
Swedish adoptees and criminality: Bohman et al. 1982

CHAPTER 5 **Once in a Lifetime: Sensitive Periods**

sensitive periods and neural circuits: Knudsen 2004
synapse number in visual cortex: Huttenlocher 1990
synapse density in frontal cortex: Huttenlocher 1979; Huttenlocher and Dabholkar 1997
synapse development in monkeys: Rakic, Bourgeois, and Goldman-Rakic 1994
energy use in children's brains: Chugani 1998
owl sound localization: Keuroghlian and Knudsen 2007
adult recovery from amblyopia: Levi 2005

CHAPTER 6 **Born Linguists**

language learning and the brain: Kuhl and Rivera-Gaxiola 2008; Gervain and Mehler 2010
sensitive periods in language learning: Johnson and Newport 1989; Newport, Bavelier, and Neville 2001

bilingual children: Werker and Byers-Heinlein 2008
children who hear more words learn language faster: Hart and Risley, 1995

CHAPTER 7 **Beautiful Dreamer**

suprachiasmatic nucleus and circadian rhythms: Welsh, Takahashi, and Kay 2010
development of sleep patterns: Roffwarg, Muzio, and Dement 1966; Dement 1974
sleep enhances neural plasticity: Frank, Issa, and Stryker 2001
sleep disorders: Garcia, Rosen, and Mahowald 2001
maternal drinking and sleep: Mennella and Gerrish 1998
getting your baby to sleep: Mennella and Gerrish 1998; Weissbluth 2003
what children dream about: Foulkes 1999; Nir and Tononi 2009
night terrors and tonsillectomy: Guilleminault et al. 2003
naps and learning: Mednick et al. 2002; Mednick and Ehrman 2006

CHAPTER 8 **It's a Girl! Gender Differences**

a phase of intense adherence to a sex role: Best and Williams 1993
statistic called *d-prime*: http://www.leeds.ac.uk/educol/documents/00002182.htm
girls have more sensitive hearing: Kei et al. 1997, d' of 0.26 for infants on a measure of peripheral auditory responsiveness used to assess hearing loss; other reports found a variety of small gender differences or none at all. Note that many of the studies cited to support this idea in popular books were done in adults, not children.
toy preference in three-year-olds: Servin, Bohlin, and Berlin 1999; many others
formation of gender identity: Maccoby 1998
monkey toy preferences: Alexander and Hines 2002 in vervet monkeys; Hassett, Siebert, and Wallen 2008 in rhesus monkeys (showed male preference for trucks; dolls were not tested)
cross-gender play and homosexuality: Bailey and Zucker 1995, meta-analysis
encouraging boyish behavior has no effect: Green 1985
boys are more active than girls: Eaton and Enns 1986, meta-analysis
boys are more physically aggressive: Card et al. 2008, meta-analysis
male monkeys engage in more rough-and-tumble play: Alexander and Hines 2002 in vervet monkeys; Wallen 2005 in rhesus monkeys; many others
CAH girls are more aggressive and more active: Pasterski et al. 2007
mental rotation in infants: Moore and Johnson 2008; Quinn and Liben 2008
mental rotation in children and adults: Voyer, Voyer, and Bryden 1995, meta-analysis
mental rotation predicts math SAT scores: Casey et al. 1995
SES and mental rotation: Levine et al. 2005
video games improve spatial skills: Subrahmanyam and Greenfield 1994, *Marble Madness* in fifth graders; Okagaki and Frensch 1994, *Tetris* in college students; De Lisi

and Cammarano 1996, *Blockout* in college students; De Lisi and Wolford 2002, *The Factory* or *Stellar 7* in third graders; Feng, Spence, and Pratt 2007, *Medal of Honor* in college students, gains lasted five months; Cherney 2008, *Antz* or *Tetris* in college students
athletes and spatial ability: Ozel, Larue, and Molinaro 2004
gender segregation: Maccoby and Jacklin 1987
girls get better grades: Linn and Kessel 1996
girls' brain volume peaks earlier: Lenroot et al. 2007
girls are better at inhibitory control: Else-Quest et al. 2006, meta-analysis
girls' advantage for language: Kovas et al. 2005
boys lag at fine motor control: Kimura 2000
writing disadvantage persists through high school: Hedges and Nowell 1995
middle-class versus poor: U.S. Department of Education, Office of Educational Research and Improvement, http://nces.ed.gov/pubs98/98041.pdf
no difference in high school math performance: Hyde et al. 2010 (more boys than girls above the ninety-ninth percentile, but only among whites, not among Asian Americans)
185 women for every 100 men with a college degree: U.S. Bureau of Labor Statistics, http://www.bls.gov/news.release/nlsyth.t01.htm
men take more time to graduate: Thomas 1981; Bank 1995
SATs and grades: Kessel and Linn 1996, meta-analysis; Leonard and Jiang 1999
girls more likely to express negative emotions: Maccoby 1998
moral reasoning differences: Jaffee and Hyde 2000, meta-analysis
differences in identifying emotions: McClure 2000, meta-analysis
risk taking: Byrnes, Miller, and Schafer 1999, meta-analysis
masturbation: Petersen and Hyde 2010, meta-analysis
self-esteem: Kling et al. 1999 (peak at fifteen to eighteen years of age), meta-analysis
body image: Feingold and Mazzella 1998, meta-analysis
body image linked to progressively thinner standards: Grabe, Ward, and Hyde 2008, meta-analysis
teasing, obesity, and anorexia: Neumark-Sztainer et al. 2007
dieting increases risk of obesity: Field and Colditz 2001

CHAPTER 9 Adolescence: It's Not Just About Sex

a time of near limitless possibility: Luciana 2010
adolescent behavior: Smetana, Campione-Barr, and Metzger 2006
brain appears nearly finished in late childhood: Caviness et al. 1996
human synapse development: Huttenlocher 1990; Glantz et al. 2007
monkey synapse development: Rakic, Bourgeois, and Goldman-Rakic 1994

gray matter and white matter development: Gogtay and Thompson 2010
brain growth and intelligence: Shaw et al. 2006a
differences in the balance between impulse and restraint: Casey, Duhoux, and Cohen 2010
Iowa Gambling Task in adolescence: Steinberg 2010
reward and the ventral striatum: Casey, Duhoux, and Cohen 2010
puberty and the adolescent brain: Sisk and Foster 2004
circadian rhythms set time to wake up and sleep: Duffy, Rimmer, and Czeisler 2001
adolescent day-night cycle: Carskadon et al. 1999; Carskadon, Acebo, and Jenni 2004
sleep duration in children and adolescents: Iglowstein et al. 2003
survey of sleep habits in girls: Frey et al. 2009
social jetlag: Wittmann et al. 2006
stress and the adolescent brain: Romeo and McEwen 2006
dopamine and novelty seeking: Spear 2010
oxytocin and parental love: Gordon et al. 2010

CHAPTER 10 Learning to See

motion: Braddick and Atkinson 2009
heritability of myopia: Hornbeak and Young 2009
increasing prevalence of myopia: Rose et al. 2001
myopia in Israel: Dayan et al. 2005
outdoor activity and myopia: Jones et al. 2007; Rose et al. 2008a; Rose et al. 2008b
contrast: Brown and Lindsey 2009
figure (what babies see): data from Salomão and Ventura 1995; Adams and Courage 2002
face processing: de Schonen et al. 2005
sensitive periods: Lewis and Maurer 2009
effects of sensory deficits: Lewis and Maurer 2009
advantage in distinguishing own- versus other-race faces: Kelly et al. 2007
corrective glasses for amblyopia: Moseley, Fielder, and Stewart 2009

CHAPTER 11 Connect with Your Baby Through Hearing and Touch

how hearing works: Kandel, Schwartz, and Jessell 2000
development of vestibular system: Nandi and Luxon 2008
prenatal risks to hearing and vestibular system: Eliot 1999
babies can hear before they're born: Hepper and Shahidullah 1994
newborns prefer mother's voice: DeCasper and Fifer 1980
newborns prefer mother's language: Mehler et al. 1988
cochlear implant timing: Harrison, Gordon, and Mount 2005
neural development of hearing and touch: Meisami and Timiras 1988, cited in Eliot 1999
processing speed: Görke 1986; Müller, Ebner, and Hömberg 1994

somatosensory cortex map: Kandel, Schwartz, and Jessell 2000
noise and hearing loss: Daniel 2007; http://www.caohc.org/updatearticles/spring07.pdf
neural pathway for pleasant touch: Olausson et al. 2002; Löken et al. 2009
for more on touch deprivation experiments in monkeys: Blum 2002

CHAPTER 12 Eat Dessert First: Flavor Preferences

newborns distinguish mother by smell: Winberg and Porter 1998
mothers distinguish newborns by smell: Kaitz et al. 1987
baby's preferences for amniotic fluid and Mom's unwashed breast: Sullivan 2000
prenatal development of olfactory brain: Schaal 1988
what toddlers eat: Fox et al. 2004
how to get kids to like broccoli: Capaldi 1996, chapter 3
organization of taste system: Yarmolinsky, Zuker, and Ryba 2009
development of taste buds: Mistretta and Bradley 1984; Witt and Reutter 1996
nucleus solitarius: Rolls 2006
morphine replacing sweet receptor: Zhao et al. 2003
learning food preferences: Mennella, Jagnow, and Beauchamp 2001
rabbits learn food preferences: Bilkó, Altbäcker, and Hudson 1994
inadequate nutrition in low-fat diets for children: Nicklas et al. 1992; Milner and Allison 1999
prevalence of eating disorders: Hoek 2006
longitudinal predictors of obesity and eating disorders: Neumark-Sztainer et al. 2007
morning sickness and salt preference: Crystal and Bernstein 1998
infant flavor preferences last for years: Mennella and Beauchamp 2002
soy or hydrolysate versus milk formula: Beauchamp and Mennella 2009
cheese versus body odor labeling: de Araujo et al. 2005

CHAPTER 13 The Best Gift You Can Give: Self-Control

preschool delay times and later success: Shoda, Mischel, and Peake 1990; Duckworth and Seligman 2005; Blair and Razza 2007
self-control of emotion; development of attention and effortful control: Bell and Deater-Deckard 2007
brain circuits for self-control: Posner and Rothbart 2007
marshmallow task strategy: Peake, Hebl, and Mischel 2002
Tools of the Mind: Diamond et al. 2007; Blair and Diamond 2008
executive function in bilingual children: Bialystok 2009; Kovács and Mehler 2009
theory of mind in bilingual children: Goetz 2003
training willpower in adults: Baumeister et al. 2006
computerized attention training: Tang and Posner 2009

CHAPTER 14 Playing for Keeps

animal play: Fagen 1974; Burghardt and Sutton-Smith 2005
brain size and play: Iwaniuk, Nelson, and Pellis 2001
octopus play: Mather and Anderson 1999
learning and stress: Joëlsa et al. 2006
video games: Gentile and Stone 2005
play in mammals: Vanderschuren, Niesink, and van Ree 1997
culture and play: Tamis-LeMonda et al. 1992

CHAPTER 15 Moving the Body and Brain Along

emotional benefits of exercise in children: Bailey 2006
cognitive benefits of exercise in children: meta-analysis Sibley and Etnier 2003; Hillman, Erickson, and Kramer 2008; Pontifex et al. 2011
cognitive benefits of exercise in adults: Hillman, Erickson, and Kramer 2008
brain changes with exercise in children: Chaddock et al. 2010; Chaddock et al. 2011
chronic traumatic encephalopathy: McKee et al. 2009
football and concussion: Guskiewicz et al. 2005; Guskiewicz et al. 2007
concussions in children: Halstead et al. 2010

CHAPTER 16 Electronic Entertainment and the Multitasking Myth

video games and attention in college students: Green and Bavelier 2003; Dye, Green, and Bavelier 2009b; Li et al. 2009
recent decline in empathy: http://www.ns.umich.edu/htdocs/releases/story.php?id=7724
video games and attention in children: Dye, Green, and Bavelier 2009a; Dye and Bavelier 2010
infants and TV: Christakis 2009; Bavelier, Green and Dye 2010
costs of switching, role of prefrontal cortex: Marois and Ivanoff 2005
distractibility in multitasking college students: Ophir, Nass, and Wagner 2009
Dora the Explorer versus *Teletubbies*: Linebarger and Walker 2005

CHAPTER 17 Nice to Meet You: Temperament

temperament and personality: Caspi 2000; Rothbart, Ahadi, and Evans 2000
high-reactive babies: Fox et al. 2005
personality traits and stability: McAdams and Olson 2010
your kids have different environments: Plomin et al. 2001
temperament and parenting in development of conscience: Kochanska and Aksan 2006
children with a specific receptor sensitive to parenting style: Sheese et al. 2007
negative feedback on antisocial behavior: Reiss and Leve 2007; Dodge and McCourt 2010

children learn from their peers: Berndt and Murphy 2002
birth order and personality: Harris 1998 (pp. 365–78); Townsend 2000
Chinese versus Canadian mothers and behavioral inhibition: Chen et al. 1998
rat grooming and pup behavior: Zhang and Meaney 2010
high-reactive monkeys are more vulnerable: Suomi 1997

CHAPTER 18 **Emotions in the Driver's Seat**

learning to recognize emotions in others: Leppänen and Nelson 2009
neural circuits involved in emotion perception: Barrett et al. 2007
left/right brain and emotions: Wager et al. 2003
self-control and empathy: Posner and Rothbart 2007
development of emotion regulation: Eisenberg, Spinrad, and Eggum 2010

CHAPTER 19 **Empathy and Theory of Mind**

theory of mind in children: Frith and Frith 2003; Saxe, Carey, and Kanwisher 2004; Singer 2006; Baillargeon, Scott, and He 2009
empathy in animals: Rice and Gainer 1962; Preston and de Waal 2002; Emery and Clayton 2009
insula: Craig 2009
older siblings and theory of mind: Ruffman et al. 1998
social brain: Adolphs 2003
mirror neurons: Rizzolatti and Sinigaglia 2010
children can't remember their past mental states: Gopnik 1993

CHAPTER 20 **Playing Nicely with Others**

parent-infant synchrony: Feldman 2007
distress when their partner stops responding: Mesman, van IJzendoorn, and Bakermans-Kranenburg 2009
maternal depression: Feldman 2007; Field, Diego, and Hernandez-Reif 2009
stranger anxiety develops: Kagen 2003
joint attention in infants: Parlade et al. 2009
attachment and day care: Friedman and Boyle 2008; Bohlin and Hagekull 2009
attachment, 5-HTT, and self-control: Kochanska, Philibert, and Barry 2009
development of peer interactions: Rubin, Bukowski, and Parker 2006
cultural effects on socialization: Chen and French 2008
mutual dislike relationships: Card 2010
conscience in children: Kochanska and Aksan 2006
social withdrawal: Rubin, Coplan, and Bowker 2009

CHAPTER 21 **Starting to Write the Life Story**

declarative versus nondeclarative memory: Squire 1987
toddlers navigate by landmarks: Wang, Hermer, and Spelke 1999
habituation in the brain: Colombo and Mitchell 2009
study habits: Rohrer and Pashler 2010
formation and breakage of synapses: Chklovskii, Mel, and Svoboda 2004
conversion to long-term memory: Squire and Zola-Morgan 1991; Wang and Morris 2010
babies forget faster than older children: Rovee-Collier and Barr 2001
memories appear to be rewritten: Wang and Morris 2010

CHAPTER 22 **Learning to Solve Problems**

beliefs about intelligence predict performance: Blackwell, Trzesniewski, and Dweck 2007
social rejection and IQ: Baumeister, Twenge, and Nuss 2002
predicting infant IQ from habituation: Kavšek 2004
heritability of IQ in poverty: Turkheimer et al. 2003
IQ in adopted children: van IJzendoorn, Juffer, and Klein Poelhuis 2005
genetic influences on intelligence and brain: Shaw 2007; Green et al. 2008; Deary, Penke, and Johnson 2010
children's brains and abstract reasoning: Ferrer, O'Hare, and Bunge 2009
Flynn effect and modern life: Dickens and Flynn 2001
interventions to improve intelligence: Buschkuehl and Jaeggi 2010
working memory training improves intelligence: Jaeggi et al. 2008

CHAPTER 23 **Take It from the Top: Music**

consonance and dissonance in infancy: Trainor, Tsang, and Cheung 2002
openness to different rhythms: Hannon and Trehub 2005
early rhythm learning: Gerry, Faux, and Trainor 2010
auditory discrimination in preschool children: Jensen and Neff 1993
rhythm perception in deaf children with cochlear implants: Nakata et al. 2006
key and harmony perception: Corrigall and Trainor, 2009; Hannon and Trainor 2007
Heschl's gyrus in musicians: Schneider et al. 2002
cognitive changes from music and drama lessons: Schellenberg 2005
practice and long-distance connections: Zatorre, Chen, and Penhune 2007
brain changes from keyboard lessons: Hyde et al. 2009
corpus callosum in children receiving musical training: Schlaug et al. 1995

CHAPTER 24 **Go Figure: Learning About Math**

Mickey doll becomes two Mickeys: Dehaene 1999

subitization, numerosity, and addition in animals: Dehaene 1999
children temporarily lose numerosity: Mehler and Bever 1967
mental number line and brain regions that process number: Nieder and Dehaene 2009
eye movements and math: Knops et al. 2009
stereotypes and performance: Wheeler and Petty 2001
performance with positive versus negative stereotypes: McGlone and Aronson 2006
Mundurukú: Pica et al. 2004
story versus equation problems: Sohn et al. 2004; Lee et al. 2007

CHAPTER 25 The Many Roads to Reading

brain activity in adults during reading: Dehaene 2009
neurons that represent objects: Desimone et al. 1984
car recognition: Gauthier et al. 2000
right inferior temporal cortex and mirror confusion: Gross and Bornstein 1978
left-right discrimination and readiness to read: Fisher, Bornstein, and Gross 1985
monkeys with damage to inferior temporal cortex: Holmes and Gross 1984
brain activity in children during reading: Turkeltaub et al. 2003
brain activity during phonological tasks: Wagner et al. 1997
reading at home: Whitehurst et al. 1988
potential causes of dyslexia: Ramus et al. 2003
survey of students at Göteborg University: Wolff and Lundberg 2002
Chinese children learn to read by writing: Tan et al. 2005
brain activity during reading in Chinese speakers: Siok et al. 2008
prevalence of dyslexia in English and Chinese speakers: Yin and Weekes 2003
Chinese dyslexics learning English: Ho and Fong 2005

CHAPTER 26 Hang in There, Baby: Stress and Resilience

children exposed to a moderate amount of stress: Power 2004
maternal separation in monkeys: Suomi 2006
maternal separation in rats: Zhang and Meaney 2010
psychological development of coping in children: Skinner and Zimmer-Gembeck 2007
overprotection of high-reactive children: Fox et al. 2005
parents regulate children's physiological stress responses: Gunnar and Quevedo 2007
effects of early life stress are similar in people: Haglund et al. 2007; Feder, Nestler, and Charney 2009
maternal effects on rat stress responses: Cameron et al. 2005; Zhang and Meaney 2010
peer-raised monkeys: Suomi 2006
longitudinal study in New Zealand: Caspi et al. 2003

serotonin transporter interaction with environmental trauma: Uher and McGuffin 2010, meta-analysis
dandelion and orchid children: Boyce and Ellis 2005; Ellis, Jackson, and Boyce 2006; related ideas in Belsky, Bakermans-Kranenburg, and van IJzendoorn 2007
one prospective study of high-stress families: Boyce et al. 1995
experimental study with rhesus monkeys: Suomi 1997

CHAPTER 27 Mind-Blindness: Autism

characteristics of autism: Frith 2003
factors that account for rise in diagnosis: Wazana, Bresnahan, and Kline 2007; Hertz-Picciotto and Delwiche 2009; Nassar et al. 2009
feral children: McCarthy 1925; Williams 2003
early signs of autism: Volkmar, Chawarska, and Klin 2005; Karmel et al. 2010
changes in amygdala, cerebellum, and neocortex in autism: Palmen et al. 2004; Amaral et al. 2008
prematurity, cerebellar injury, and autism: Limperopoulos 2010
cerebellum and tickling: Blakemore, Wolpert, and Frith 1998
Temple Grandin quote: Grandin 1995
L. H. Willey quote: Willey 1999
half of children receiving intensive behavioral therapy can enter regular education: Sallows and Graupner 2005
UCLA model of therapy: Cohen, Amerine-Dickens, and Smith 2006
combining intensive behavioral therapy with other approaches is less effective: Howard et al. 2005
therapy for Catherine: Green, Brennan, and Fein 2002
very early intervention: Dawson et al. 2010
Wakefield vaccine fraud: Deer 2011

CHAPTER 28 Old Genes Meet the Modern World: ADHD

estimates of ADHD prevalence: Castellanos and Tannock 2002; Elia et al. 2010
ADHD susceptibility genes: Elia et al. 2010
white and gray matter growth in ADHD: Shaw et al. 2006b
discovery of Ritalin: Shorter 2008
EEG rhythms in ADHD: Barry, Clarke, and Johnstone 2003
no evidence for sweeping claims about movement exercises: Bishop 2007
warning signs of quackery: Hyman and Levy 2000; Jacobson, Foxx, and Mulick 2005
meta-analysis of neurofeedback studies: Arns et al. 2009
functional imaging is not a reliable diagnostic tool: Bush, Valera, and Seidman 2005

differences in caudate nucleus, basal ganglia: Mink 1996; Redgrave, Prescott, and Gurney 1999; Tripp and Wickens 2009

Amen Clinics evaluated: http://www.quackwatch.org/06ResearchProjects/amen.html and http://www.salon.com/life/mind_reader/2008/05/12/daniel_amen/.

Amen's books offer what most psychiatrists would do: Leuchter 2009

Amen has declined the opportunity to test his diagnostic tools: Adinoff and Devous 2010

by age eighteen, symptoms have subsided: Mannuzza et al. 1991

Ritalin does not increase risk of substance abuse: Mannuzza et al. 2008

brain differences reflect a delay in development: Shaw et al. 2007; Castellanos et al. 2002

CHAPTER 29 Catch Your Child Being Good: Behavior Modification

modifying children's behavior: Strand 2000; Kazdin and Rotella 2009

ratio of positive to negative interactions in marriage: Gottman and Silver 1998

extinction learning and the brain: Quirk and Mueller 2008

self-esteem outcomes: Baumeister et al. 2003

praise and motivation: Henderlong and Lepper 2002

CHAPTER 30 A Tough Road to Travel: Growing Up in Poverty

relationship between SES and health: Sapolsky 2004; Dow and Rehkopf 2010

stress and the health effects of SES: Sapolsky 2005; McEwen and Gianaros 2010; but see Matthews, Gallo, and Taylor 2010 for a contrary view

brain effects of deprivation: Hackman and Farah 2009; McEwen and Gianaros 2010

language environment provided by low-SES parents: Hart and Risley 1995

adopted children and biological parents' SES: Rowe and Rodgers 1997

likely relationship between poverty and achievement is a vicious cycle: Conger and Donnellan 2007

exposure to toxins: Evans 2006

epidemiology interpretation: Taubes 2007

nun study: Snowdon et al. 1989

people whose SES improves later in life: Cohen et al. 2010

children who move out of poverty: Costello et al. 2003; Kawachi, Adler, and Dow 2010

intensive preschool enrichment: Knudsen et al. 2006

GLOSSARY

For further introduction to technical terms: Bear, Connors, and Paradiso 2006

REFERENCES

Adams, R. J., and Courage, M. L. Using a single test to measure human contrast sensitivity from early childhood to maturity. *Vision Research* 42:1205–10 (2002).

Adinoff, B., and Devous, M. Scientifically unfounded claims in diagnosing and treating patients. *American Journal of Psychiatry* 167:598 (2010).

Adolph, K. E., Karasik, L. B., and Tamis-LeMonda, C. S. Motor skill. In *Handbook of Cultural Developmental Science*, ed. M. H., Bornstein, 61–88. New York: Psychology Press, 2009.

Adolphs, R. Cognitive neuroscience of human social behaviour. *Nature Reviews Neuroscience* 4:165–78 (2003).

Alexander, G. M., and Hines, M. Sex differences in response to children's toys in nonhuman primates (Cercopithecus aethiops sabaeus). *Evolution and Human Behavior* 23:467–79 (2002).

Amaral, D. G., Schumann, C. M., and Nordahl, C. W. Neuroanatomy of autism. *Trends in Neurosciences* 31:137–45 (2008).

Ananth, C. V., Joseph, K. S., Oyelese, Y., Demissie, K., and Vintzileos, A. M. Trends in preterm birth and perinatal mortality among singletons: United States, 1989 through 2000. *Obstetrics and Gynecology* 105:1084–91 (2005).

Arai, J. A., Li, S., Hartley, D. M., and Feig, L. A. Transgenerational rescue of a genetic defect in long-term potentiation and memory formation by juvenile enrichment. *Journal of Neuroscience* 29:1496–1502 (2009).

Arns, M., de Ridder, S., Strehl, U., Breteler, M., and Coenen A. Efficacy of neurofeedback treatment in ADHD: The effects on attention, impulsivity, and hyperactivity: a meta-analysis. *Clinical EEG and Neuroscience* 40:180–89 (2009).

Bailey, J. M., and Zucker, K. J. Childhood sex-typed behavior and sexual orientation: A conceptual analysis and quantitative review. *Developmental Psychology* 31:43–55 (1995).

Bailey, R. Physical education and sport in schools: A review of benefits and outcomes. *Journal of School Health* 76:397–401 (2006).

Baillargeon, R., Scott, R. M., and He, Z. False-belief understanding in infants. *Trends in Cognitive Science* 14:110–18 (2009).

Baillargeon, R., Spelke, E. S., and Wasserman, S. Object permanence in five-month-old infants. *Cognition* 20:191–208 (1985).

Baillargeon, R., and Wang, S. Event categorization in infancy. *Trends in Cognitive Science* 6:85–93 (2002).

Bank, B. J. Gendered accounts: Undergraduates explain why they seek their bachelor's degree. *Sex Roles* 32:527–44 (1995).

Barrett, L. F., Mesquita, B., Ochsner, K. N., and Gross, J. J. The experience of emotion. *Annual Review of Psychology* 58:373–403 (2007).

Barry, R. J., Clarke, A. R., and Johnstone,

S. J. A review of electrophysiology in attention-deficit/hyperactivity disorder: I. Qualitative and quantitative electroencephalography. *Clinical Neurophysiology* 114:171–83 (2003).

Baumeister, R. F., Campbell, J. D., Krueger, J. I., and Vohs, K. D. Does high self-esteem cause better performance, interpersonal success, happiness, or healthier lifestyles? *Psychological Science in the Public Interest* 4:1–44 (2003).

Baumeister, R. F., Gailliot, M., DeWall, C. N., and Oaten, M. Self-regulation and personality: How interventions increase regulatory success, and how depletion moderates the effects of traits on behavior. *Journal of Personality* 74:1773–1802 (2006).

Baumeister, R. F., Twenge, J. M., and Nuss, C. K. Effects of social exclusion on cognitive processes: Anticipated aloneness reduces intelligent thought. *Journal of Personality and Social Psychology* 83:817–27 (2002).

Bavelier, D., Green, C. S., and Dye, M. W. G. Children, wired: For better and for worse. *Neuron* 67:692–701 (2010).

Bear, M. F., Connors, B. W., and Paradiso, M. A. *Neuroscience: Exploring the Brain.* 3rd edition. Philadelphia: Lippincott, 2006.

Beauchamp, G. K., and Mennella, J. A. Early flavor learning and its impact on later feeding behavior. *Journal of Pediatric Gastroenterology and Nutrition* 48:S25–S30 (2009).

Bell, M. A., and Deater-Deckard, K. Biological systems and the development of self-regulation: Integrating behavior, genetics, and psychophysiology. *Journal of Developmental and Behavioral Pediatrics* 28:409–20 (2007).

Belsky, J., Bakermans-Kranenburg, M. J., and van IJzendoorn, M. H. For better and for worse: Differential susceptibility to environmental influences. *Current Directions in Psychological Science* 16:300–304 (2007).

Berndt, T. J., and Murphy, L. M. Influences of friends and friendships: Myths, truths, and research recommendations. *Advances in Child Development and Behavior* 30:275–310 (2002).

Best, D. L., and Williams, J. E. A cross-cultural viewpoint. In *The Psychology of Gender*, ed. A. E. Beall and R. J. Sternberg, 215–48. New York: Guilford, 1993.

Bialystok, E. Bilingualism: The good, the bad, and the indifferent. *Bilingualism: Language and Cognition* 12:3–11 (2009).

Bilkó, A., Altbäcker, V., and Hudson, R. Transmission of food preference in the rabbit: The means of information transfer. *Physiology and Behavior* 56:907–12 (1994).

Bishop, D. V. M. Curing dyslexia and attention-deficit hyperactivity disorder by training motor co-ordination: Miracle or myth? *Journal of Paediatrics and Child Health* 43:653–55 (2007).

Blackwell, L. S., Trzesniewski, K. H., and Dweck, C. S. Implicit theories of intelligence predict achievement across an adolescent transition: A longitudinal study and an intervention. *Child Development* 78:246–63 (2007).

Blair, C., and Diamond, A. Biological processes in prevention and intervention: The promotion of self-regulation as a means of preventing school failure. *Development and Psychopathology* 20:899–911 (2008).

Blair, C., and Razza, R. P. Relating effortful control, executive function, and false belief understanding to emerging math and literacy ability in kindergarten. *Child Development* 78:647–63 (2007).

Blakemore, S. J., Wolpert, D. M., and Frith, C. D. Central cancellation of self-produced tickle sensation. *Nature Neuroscience* 1:635–40 (1998).

Bloom, P. *Descartes' Baby: How the Science of Child Development Explains What Makes Us Human.* New York: Basic Books, 2004.

Blum, D. *Love at Goon Park: Harry Harlow and the Science of Affection.* New York: Perseus, 2002.

Bohlin, G., and Hagekull, B. Socio-emotional development: From infancy to young adulthood. *Scandinavian Journal of Psychology* 50:592–601 (2009).

Bohman, M., Cloninger, R., Sigvardsson, S., and von Knorring, A.-L. Predisposition to petty criminality in Swedish adoptees. I. Genetic and environmental heterogeneity. *Archives of General Psychiatry* 39:1233–41 (1982).

Boyce, W. T., Chesney, M., Alkon-Leonard, A., Tschann, J., Adams, S., Chesterman, B., Cohen, F., Kaiser, P., Folkman, S., and Wara, D. Psychobiologic reactivity to stress and childhood respiratory illnesses: Results of two prospective studies. *Psychosomatic Medicine* 57:411–22 (1995).

Boyce, W. T., and Ellis, B. J. Biological sensitivity to context: I. An evolutionary-developmental theory of the origins and functions of stress reactivity. *Development and Psychopathology* 17:271–301 (2005).

Braddick, O. J., and Atkinson, J. Infants' sensitivity to motion and temporal change. *Optometry and Vision Science* 86:577–82 (2009).

Brown, A. M., and Lindsey, D. T. Contrast insensitivity: The critical immaturity in infant visual performance. *Optometry and Vision Science* 86:572–76 (2009).

Burghardt, G. M., and Sutton-Smith, B. *Genesis of Animal Play: Testing the Limits.* Cambridge, MA: MIT Press, 2005.

Buschkuehl, M., and Jaeggi, S. M. Improving intelligence: A literature review. *Swiss Medical Weekly* 140:266–72 (2010).

Bush, G., Valera, E. M., and Seidman, L. J. Functional neuroimaging of attention-deficit/hyperactivity disorder: A review and suggested future directions. *Biological Psychiatry* 57:1273–84 (2005).

Byrnes, J. P., Miller, D. C., and Schafer, W. D. Gender differences in risk taking: A meta-analysis. *Psychological Bulletin* 125:367–83 (1999).

Cameron, N. M., Champagne, F. A., Parent, C., Fish, E. W., Ozaki-Kuroda, K., and Meaney, M. J. The programming of individual differences in defensive responses and reproductive strategies in the rat through variations in maternal care. *Neuroscience and Biobehavioral Reviews* 29:843–65 (2005).

Capaldi, E. D., ed. *Why We Eat What We Eat: The Psychology of Eating.* Washington, DC: American Psychological Association, 1996.

Card, N. A. Antipathetic relationships in child and adolescent development: A meta-analytic review and recommendations for an emerging area of study. *Developmental Psychology* 46:516–29 (2010).

Card, N. A., Stucky, B. D., Sawalani, G. M., and Little, T. D. Direct and indirect aggression during childhood and adolescence: A meta-analytic review of gender differences, intercorrelations, and relations to maladjustment. *Child Development* 79:1185–1229 (2008).

Carskadon, M. A., Acebo, C., and Jenni, O. G. Regulation of adolescent sleep: Implications for behavior. *Annals of the New York Academy of Sciences* 1021:276–91 (2004).

Carskadon, M. A., Labyak, S. E., Acebo, C., and Seifer, R. Intrinsic circadian period of adolescent humans measured in conditions of forced desynchrony. *Neuroscience Letters* 260:129–32 (1999).

Casey, B. J., Duhoux, S., and Cohen, M. M. Adolescence: What do transmission, transition, and translation have to do with it? *Neuron* 67:749–60 (2010).

Casey, M. B., Nuttall, R., Pezaris, E., and Benbow, C. P. The influence of spatial ability on gender differences in mathematics college entrance test scores across diverse samples. *Developmental Psychology* 31:697–705 (1995).

Caspi, A. The child is father of the man: Personality continuities from childhood to adulthood. *Journal of Personality and Social Psychology* 78:158–72 (2000).

Caspi, A., Sugden, K., Moffitt, T. E., Taylor, A., Craig, I. W., Harrington, H., McClay, J., Mill, J., Martin, J., Braithwaite, A., and Poulton, R. Influence of life stress on depression: Moderation by a polymorphism in the 5-HTT gene. *Science* 301:386–89 (2003).

Castellanos, F. X., Lee, P. P., Sharp, W., Jeffries, N. O., Greenstein, D. K., Clasen, L. S., Blumenthal, J. D., James, R. S., Evens, C. L., Walter, J. M., Zijdenbos, A., Evans, A. C., Giedd, J. N., and Rapoport, J. L. Developmental trajectories of brain volume abnormalities in children and adolescents with attention-deficit/hyperactivity disorder. *Journal of the American Medical Association* 288:1740–48 (2002).

Castellanos, F. X., and Tannock, R. Neuroscience of attention-deficit/hyperactivity disorder: The search for endophenotypes. *Nature Reviews Neuroscience* 3:617–28 (2002).

Caviness, V. S., Jr., Kennedy, D. N., Richelme, C., Rademacher J., and Filipek, P. A. The human brain age 7–11 years: A volumetric analysis based on magnetic resonance images. *Cerebral Cortex* 6:726–36 (1996).

Chaddock, L., Erickson, K. I., Prakash, R. S., VanPatter, M., Voss, M. W., Pontifex, M. B., Raine, L. B., Hillman, C. H., and Kramer, A. F. Basal ganglia volume is associated with aerobic fitness in preadolescent children. *Developmental Neuroscience* 32:249–56 (2010).

Chaddock, L., Hillman, C. H., Buck, S. M., and Cohen, N. J. Aerobic fitness and executive control of relational memory in preadolescent children. *Medicine and Science in Sports and Exercise* 43:344–49 (2011).

Chen, X., and French, D. C. Children's social competence in cultural context. *Annual Review of Psychology* 59:591–616 (2008).

Chen, X., Hastings, P., Rubin, K. H., Chen, H., Cen, G., and Stewart, S. L. Childrearing attitudes and behavioral inhibition in Chinese and Canadian toddlers: A cross-cultural study. *Developmental Psychology* 23:677–86 (1998).

Cherney, I. D. Mom, let me play more computer games: They improve my mental rotation skills. *Sex Roles* 59:776–86 (2008).

Chklovskii, D. B., Mel, B. W., and Svoboda, K. Cortical rewiring and information storage. *Nature* 431:782–88 (2004).

Christakis, D. A. The effects of infant media usage: What do we know and what should we learn? *Acta Paediatrica* 98:8–16 (2009).

Chugani, H. T. A critical period of brain development: Studies of cerebral glucose utilization with PET. *Preventive Medicine* 27:184–88 (1998).

Cohen, H., Amerine-Dickens, M., and Smith, T. Early intensive behavioral treatment: Replication of the UCLA model in a community setting. *Journal of Developmental and Behavioral Pediatrics* 27:S145–55 (2006).

Cohen, S., Janicki-Deverts, D., Chen, E., and Matthews, K. A. Childhood socioeconomic status and adult health. *Annals of the New York Academy of Sciences* 1186:37–55 (2010).

Colombo, J. The development of visual attention in infancy. *Annual Review of Psychology* 52:337–67 (2001).

Colombo, J., and Mitchell, D. W. Infant visual habituation. *Neurobiology of Learning and Memory* 92:225–34 (2009).

Conger, R. D., and Donnellan, M. B. An interactionist perspective on the socioeconomic context of human development. *Annual Review of Psychology* 58:175–99 (2007).

Corrigall, K. A., and Trainor, L. J. Effects of musical training on key and harmony perception. *Annals of the New York Academy of Sciences* 1169:164–68 (2009).

Costello, E. J., Compton, S. N., Keeler, G., and Angold, A. Relationships between poverty and psychopathology: A natural experiment. *Journal of the American Medical Association* 290:2023–29 (2003).

Craig, A. D. How do you feel—now? The anterior insula and human awareness. *Nature Reviews Neuroscience* 10:59–70 (2009).

Crystal, S., and Bernstein, I. L. Infant salt preference and mother's morning sickness. *Appetite* 30:297–307 (1998).

Daniel, E. Noise and hearing loss: A review. *Journal of School Health* 77:225–31 (2007).

Dawson, G., Rogers, S., Munson, J., Smith, M., Winter, J., Greenson, J., Donaldson, A., and Varley, J. Randomized, controlled trial of an intervention for toddlers with autism: The Early Start Denver Model. *Pediatrics* 125:e17–23 (2010).

Dayan, Y. B., Levin, A., Morad, Y., Grotto, I., Ben-David, R., Goldberg, A., Onn, E., Avni, I., Levi, Y., and Benyamini, O. G. The changing prevalence of myopia in young adults: A 13-year series of population-based prevalence surveys. *Investigative Ophthalmology and Visual Science* 46:2760–65 (2005).

de Araujo, I. E., Rolls, E. T., Valazco, M. I., Margot, C., and Cayeux, I. Cognitive modulation of olfactory processing. *Neuron* 46:671–79 (2005).

De Lisi, R., and Cammarano, D. M. Computer experience and gender differences in undergraduate mental rotation performance. *Computers in Human Behavior* 12:351–61 (1996).

De Lisi, R., and Wolford, J. L. Improving children's mental rotation accuracy with computer game playing. *Jour-*

nal of Genetic Psychology 163:272–82 (2002).

de Schonen, S., Mancini, J., Camps, R., Maes, E., and Laurent, A. Early brain lesions and face-processing development. *Developmental Psychobiology* 46:184–208 (2005).

Deary, I. J., Penke, L., and Johnson, W. The neuroscience of human intelligence differences. *Nature Reviews Neuroscience* 11:201–11 (2010).

DeCasper, A., and Fifer, W. Of human bonding: Newborns prefer their mothers' voices. *Science* 12:305–17 (1980).

Deer, B. How the case against the MMR vaccine was fixed. *British Medical Journal* 342:c5347 (2011).

Dehaene, S. *The Number Sense: How the Mind Creates Mathematics*. New York: Oxford University Press, 1999.

———. *Reading in the Brain: The Science and Evolution of a Human Invention*. New York: Viking, 2009.

Dement, W. C. *Some Must Watch While Some Must Sleep*. San Francisco: W. H. Freeman, 1974.

Der, G., Batty, G. D., and Deary, I. J. Effect of breast feeding on intelligence in children: Prospective study, sibling pairs analysis, and meta-analysis. *British Medical Journal* 333:945–50 (2006).

Desimone, R., Albright, T. G., Gross, C. G., and Bruce, C. Stimulus-selective properties of inferior temporal neurons in the macaque. *Journal of Neuroscience* 4:2051–62 (1984).

Diamond, A., Barnett, W. S., Thomas, J., and Munro, S. Preschool program improves cognitive control. *Science* 318:1387–88 (2007).

Dickens, W. T., and Flynn, J. R. Heritability estimates versus large environmental effects: The IQ paradox resolved. *Psychological Review* 108:346–69 (2001).

Dodge, K. A., and McCourt, S. N. Translating models of antisocial behavioral development into efficacious intervention policy to prevent adolescent violence. *Developmental Psychobiology* 52:277–85 (2010).

Dow, W. H., and Rehkopf, D. H. Socioeconomic gradients in health in international and historical context. *Annals of the New York Academy of Sciences* 1186:24–36 (2010).

Duckworth, A. L., and Seligman, M. E. P. Self-discipline outdoes IQ in predicting academic performance of adolescents. *Psychological Science* 16:939–44 (2005).

Duffy, J. F., Rimmer, D. W., and Czeisler, C. A. Association of intrinsic circadian period with morningness-eveningness, usual wake time, and circadian phase. *Behavioral Neuroscience* 115:895–99 (2001).

Dye, M. W. G., and Bavelier, D. Differential development of visual attention skills in school-age children. *Vision Research* 50:452–59 (2010).

Dye, M. W. G., Green, C. S., and Bavelier, D. The development of attention skills in action video game players. *Neuropsychologia* 47:1780–89 (2009a).

———. Increased speed of processing with action video games. *Current Directions in Psychological Science* 18:321–26 (2009b).

Eaton, W. O., and Enns, L. R. Sex differences in human motor activity level. *Psychological Bulletin* 100:19–28 (1986).

Eisenberg, N., Spinrad, T. L., and Eggum, N. D. Emotion-related self-regulation and its relation to children's maladjustment. *Annual Review of Clinical Psychology* 6:495–525 (2010).

Elia, J., Gai, X., Xie, H. M., Perin, J. C., Geiger, E., Glessner, J. T., D'arcy, M., Deberardinis, R., Frackelton, E., Kim, C., Lantieri, F., Muganga, B. M., Wang, L., Takeda, T., Rappaport, E. F., Grant, S. F., Berrettini, W., Devoto, M., Shaikh, T. H., Hakonarson, H., and White, P. S. Rare structural variants found in attention-deficit hyperactivity disorder are preferentially associated with neurodevelopmental genes. *Molecular Psychiatry* 15:637–46 (2010).

Eliot, L. *Pink Brain, Blue Brain: How Small Differences Grow Into Troublesome Gaps and What We Can Do About It.* New York: Houghton Mifflin, 2010.

———. *What's Going on in There? How the Brain and Mind Develop in the First Five Years of Life.* New York: Bantam, 1999.

Ellis, B. J., Jackson, J. J., and Boyce, W. T. The stress response systems: Universality and adaptive individual differences. *Developmental Review* 26:175–212 (2006).

Else-Quest, N. M., Hyde, J. S., Goldsmith, H. H., and Van Hulle, C. Gender differences in temperament: A meta-analysis. *Psychological Bulletin* 132:33–72 (2006).

Emery, N. J., and Clayton, N. S. Comparative social cognition. *Annual Review of Psychology* 60:87–113 (2009).

Engle, W. A., and Kominiarek, M. A. Late preterm infants, early term infants, and timing of elective deliveries. *Clinics in Perinatology* 35:325–41 (2008).

Evans, G. W. Child development and the physical environment. *Annual Review of Psychology* 57:423–51 (2006).

Evenhouse, E., and Reilly, S. Improved estimates of the benefits of breastfeeding using sibling comparisons to reduce selection bias. *Health Services Research* 40:1781–1802 (2005).

Fagen, R. Selective and evolutionary aspects of animal play. *American Naturalist* 108:850–58 (1974).

Feder, A., Nestler, E. J., and Charney, D. S. Psychobiology and molecular genetics of resilience. *Nature Reviews Neuroscience* 10:446–57 (2009).

Feingold, A., and Mazzella, R. Gender differences in body image are increasing. *Psychological Science* 9:190–95 (1998).

Feldman, R. Parent-infant synchrony and the construction of shared timing; physiological precursors, developmental outcomes, and risk conditions. *Journal of Child Psychology and Psychiatry* 48:329–54 (2007).

Feng, J., Spence, I., and Pratt, J. Playing an action video game reduces gender differences in spatial cognition. *Psychological Science* 18:850–55 (2007).

Fernald, A. Meaningful melodies in mother's speech to infants. In *Nonverbal Vocal Communication*, ed. H. Papousek, U. Jurgens, and M. Papousek, 262–82. Cambridge, UK: Cambridge University Press, 1992.

Ferrer, E., O'Hare, E. D., and Bunge, S. A. Fluid reasoning and the developing brain. *Frontiers in Neuroscience* 3:46–51 (2009).

Field, A. E., and Colditz, G. A. Frequent dieting and the development of obesity among children and adolescents. *Nutrition* 17:355–56 (2001).

Field, T., Diego, M., and Hernandez-Reif, M. Depressed mothers' infants are less responsive to faces and voices. *Infant*

Behavior and Development 32:239–44 (2009).

Fisher, C. B., Bornstein, M. H., and Gross, C. G. Left-right coding and skills related to beginning reading. *Developmental and Behavioral Pediatrics* 6:279–83 (1985).

Foulkes, D. *Children's Dreaming and the Development of Consciousness*. Cambridge, MA: Harvard University Press, 1999.

Fox, M. K., Pac, S., Devaney, B., and Jankowski, L. Feeding infants and toddlers study: What foods are infants and toddlers eating? *Journal of the American Dietetic Association* 104:s22–s30 (2004).

Fox, N. A., Henderson, H. A., Marshall, P. J., Nichols, K. E., and Ghera, M. M. Behavioral inhibition: Linking biology and behavior within a developmental framework. *Annual Review of Psychology* 56:235–62 (2005).

Frank, M. G., Issa, N. P., and Stryker, M. P. Sleep enhances plasticity in the developing visual cortex. *Neuron* 30:275–87 (2001).

Frey, S., Balu, S., Greusing, S., Rothen, N., and Cajochen, C. Consequences of the timing of menarche on female adolescent sleep phase preference. *PLoS ONE* 4:e5217 (2009).

Friedman, S. L., and Boyle, D. E. Attachment in US children experiencing nonmaternal care in the early 1990s. *Attachment and Human Development* 10:225–61 (2008).

Frith, U. *Autism: Explaining the Enigma*. Oxford, UK: Wiley-Blackwell, 2003.

Frith, U., and Frith, C. D. Development and neurophysiology of mentalizing. *Philosophical Transactions of the Royal Society of London B* 358:459–73 (2003).

Fuchs, K., and Wapner, R. Elective Cesarean section and induction and their impact on late preterm births. *Clinics in Perinatology* 33:793–801 (2006).

Garcia, J., Rosen, G., and Mahowald, M. Circadian rhythms and circadian rhythm disorders in children and adolescents. *Seminars in Pediatric Neurology* 8:229–40 (2001).

Gauthier, I., Skudlarski, P., Gore, J. C., and Anderson, A. W. Expertise with cars and birds recruits brain areas involved in face recognition. *Nature Neuroscience* 3:191–97 (2000).

Gentile, D. A., and Stone, W. Violent video game effects on children and adolescents. A review of the literature. *Minerva Pediatrica* 57:337–58 (2005).

Gentile, S. Drug treatment for mood disorders in pregnancy. *Current Opinion in Psychiatry* 24:34-40 (2011).

Gergely, G., Bekkering, H., and Király, I. Rational imitation in preverbal infants. *Nature* 415:755 (2002).

Gergely, G., Nádasdy, Z., Csibra, G., and Bíró, S. Taking the intentional stance at 12 months of age. *Cognition* 56:165–93 (1995).

Gerry, D. W., Faux, A. L., and Trainor, L. J. Effects of Kindermusik training on infants' rhythmic enculturation. *Developmental Science* 13:545–51 (2010).

Gervain, J., and Mehler, J. Speech perceptions and language acquisition in the first year of life. *Annual Review of Psychology* 61:191–218 (2010).

Glantz, L. A., Gilmore, J. H., Hamer, R. M., Lieberman, J. A., and Jarskog, L. F. Synaptophysin and postsynaptic density protein 95 in the human prefrontal cortex from mid-gestation into early adulthood. *Neuroscience* 149:582–91 (2007).

Goetz, P. J. The effects of bilingualism on

theory of mind development. *Bilingualism: Language and Cognition* 6:1–15 (2003).

Gogtay, N., and Thompson, P. M. Mapping gray matter development: Implications for typical development and vulnerability to psychopathology. *Brain and Cognition* 72:6–15 (2010).

Goldenberg, R. L., Culhane, J. F., Iams, J. D., and Romero, R. Epidemiology and causes of preterm birth. *Lancet* 371: 75–84 (2008).

Goodkin, F. The development of mature patterns of head-eye coordination in the human infant. *Early Human Development* 4:373–86 (1980).

Gopnik, A. How we know our minds: The illusion of first-person knowledge of intentionality. *Behavioral and Brain Sciences* 16:1–14 (1993).

Gopnik, A. *The Philosophical Baby: What Children's Minds Tell Us About Truth, Love, and the Meaning of Life.* New York: Farrar, Straus and Giroux, 2009.

Gopnik, A., Meltzoff, A. N., and Kuhl, P. K. *The Scientist in the Crib: What Early Learning Tells Us About the Mind.* New York: HarperCollins, 1999.

Gopnik, A. S., and Choi, S. Do linguistic differences lead to cognitive differences? A cross-linguistic study of semantic and cognitive development. *First Language* 10:199–215 (1990).

Gordon, I., Zagoory-Sharon, O., Leckman, J. F., and Feldman, R. Oxytocin, cortisol, and triadic family interactions. *Physiology and Behavior* 101:679–84 (2010).

Görke, W. Somatosensory evoked cortical potentials indicating impaired motor development in infancy. *Developmental Medicine and Child Neurology* 28:633–41 (1986).

Gottman, J. M., and Silver, N. *Why Marriages Succeed or Fail: And How You Can Make Yours Last.* New York: Simon and Schuster, 1998.

Grabe, S., Ward, L. M., and Hyde, J. S. The role of the media in body image concerns among women: A meta-analysis of experimental and correlational studies. *Psychological Bulletin* 134:460–76 (2008).

Grandin, T. *Thinking in Pictures: And Other Reports from My Life with Autism.* New York: Doubleday, 1995.

Green, A. E., Munafò, M. R., DeYoung, C. G., Fossella, J. A., Fan, J., and Gray, J. R. Using genetic data in cognitive neuroscience: From growing pains to genuine insights. *Nature Reviews Neuroscience* 9:710–20 (2008).

Green, C. S., and Bavelier, D. Action video game modifies visual selective attention. *Nature* 423:534–37 (2003).

Green, G., Brennan, L. C., and Fein, D. Intensive behavioral treatment for a toddler at high risk for autism. *Behavior Modification* 26:69–102 (2002).

Green, R. Gender identity in childhood and later sexual orientation: Follow-up of 78 males. *American Journal of Psychiatry* 142:339–41 (1985).

Gross, C. G., and Bornstein, M. H. Left and right in science and art. *Leonardo* 11:29–38 (1978). Reprinted in Gross, *A Hole in the Head: More Tales in the History of Neuroscience.* Cambridge, MA: MIT Press, 2009.

Guilleminault, D., Palombini, L., Pelayo, R., and Chervin, R. D. Sleepwalking and sleep terrors in prepubertal children: What triggers them? *Pediatrics* 111:e17–25 (2003).

Gunnar, M., and Quevedo, K. The neurobiology of stress and

development. *Annual Review of Psychology* 58:145–73 (2007).

Guskiewicz, K. M., Marshall, S. W., Bailes, J., McCrea, M., Cantu, R. C., Randolph, C., and Jordan, B. D. Late-life cognitive impairment in retired professional football players. *Neurosurgery* 57:719–26 (2005).

Guskiewicz, K. M., Marshall, S. W., Bailes, J., McCrea, M., Harding, H. P., Jr., Matthews, A., Mihalik, J. R., and Cantu, R. C. Recurrent concussion and risk of depression in retired professional football players. *Medicine and Science in Sports and Exercise* 39:903–9 (2007).

Hackman, D. A., and Farah, M. J. Socioeconomic status and the developing brain. *Trends in Cognitive Sciences* 13:65–73 (2009).

Haglund, M. E. M., Nestadt, P. S., Cooper, N. S., Southwick, S. M., and Charney, D. S. Psychobiological mechanisms of resilience: Relevance to prevention and treatment of stress-related psychopathology. *Development and Psychopathology* 19:889–920 (2007).

Halstead, M. E., Walter, K. D., and the Council on Sports Medicine and Fitness. Clinical report—Sport-related concussion in children and adolescents. *Pediatrics* 126:597–615 (2010).

Hannon, E. E., and Trainor, L. J. Music acquisition: Effects of enculturation and formal training on development. *Trends in Cognitive Sciences* 11:466–72 (2007).

Hannon, E. E., and Trehub, S. E. Metrical categories in infancy and adulthood. *Psychological Science* 16:48–55 (2005).

Harris, J. R. The Nurture Assumption: *Why Children Turn Out the Way They Do.* New York: Simon and Schuster, 1998.

Harrison, R. V., Gordon, K. A., and Mount, R. J. Is there a critical period for cochlear implantation in congenitally deaf children? Analyses of hearing and speech perception performance after implantation. *Developmental Psychobiology* 46:252–61 (2005).

Hart, B., and Risley, T. R. *Meaningful Differences in the Everyday Experience of Young American Children.* Baltimore: Paul H. Brookes, 1995.

Hassett, J. M., Siebert, E. R., and Wallen, K. Sex differences in rhesus monkey toy preferences parallel those of children. *Hormones and Behavior* 54:359–64 (2008).

Hedges, L. V., and Nowell, A. Sex differences in mental test scores, variability, and numbers of high-scoring individuals. *Science* 269:41–45 (1995).

Henderlong, J., and Lepper, M. R. The effects of praise on children's intrinsic motivation: A review and synthesis. *Psychological Bulletin* 128:774–95 (2002).

Hepper, P. G., and Shahidullah, B. S. Development of fetal hearing. *Archives of Disease in Childhood* 71:81–87 (1994).

Hertz-Picciotto, I., and Delwiche, L. The rise in autism and the role of age at diagnosis. *Epidemiology* 20:84–90 (2009).

Hibbeln, J. R., Davis, J. M., Steer, C., Emmett, P., Rogers, I., Williams, C., and Golding, J. Maternal seafood consumption in pregnancy and neurodevelopmental outcomes in childhood (ALSPAC study): An observational cohort study. *Lancet* 369:578–85 (2007).

Hillman, C. H., Erickson, K. I., and Kramer, A. F. Be smart, exercise your heart: Exercise effects on brain and cognition. *Nature Reviews Neuroscience* 9:58–65 (2008).

Ho, C. S.-H., and Fong, K.-M. Do Chinese dyslexic children have difficulties learning English as a second language? *Journal of Psycholinguistic Research* 34:603–19 (2005).

Hoek, H. W. Incidence, prevalence and mortality of anorexia nervosa and other eating disorders. *Current Opinion in Psychiatry* 19:389–94 (2006).

Holden, C., and Mace, R. Phylogenetic analysis of the evolution of lactose digestion in adults. *Human Biology* 69:605–28 (1997).

Holmes, E. J., and Gross, C. G. Effects of inferior temporal lesions on discrimination of stimuli differing in orientation. *Journal of Neuroscience* 4:3063–68 (1984).

Hood, B. M., Willen, J. D., and Driver, J. Adult's eyes trigger shifts of visual attention in human infants. *Psychological Science* 9:131–34 (1998).

Hornbeak, D. M., and Young, T. L. Myopia genetics: A review of current research and emerging trends. *Current Opinion in Ophthalmology* 20:356–62 (2009).

Howard, J. S., Sparkman, C. R., Cohen, H. G., and Stanislaw, H. A comparison of intensive behavior analytic and eclectic treatments for young children with autism. *Research in Developmental Disabilities* 26:359–83 (2005).

Huttenlocher, P. R. Morphometric study of human cerebral cortex development. *Neuropsychologia* 28:517–27 (1990).

———. Synaptic density in human frontal cortex—developmental changes and effects of aging. *Brain Research* 163:195–205 (1979).

Huttenlocher, P. R., and Dabholkar, A. S. Regional differences in synaptogenesis in human cerebral cortex. *Journal of Comparative Neurology* 387:167–78 (1997).

Hyde, J. S., Lindberg, S. M., Linn, M. C., Ellis, A. B., and Williams, C. C. Gender similarities characterize math performance. *Science* 321:494–95 (2010).

Hyde, K. L., Lerch, J., Norton, A., Forgeard, M., Winner, E., Evans, A. C., and Schlaug, G. Musical training shapes structural brain development. *Journal of Neuroscience* 29:3019–25 (2009).

Hyman, S. L., and Levy, S. E. Autism spectrum disorders: When traditional medicine is not enough. *Contemporary Pediatrics* 17:101–16 (2000).

Iglowstein, I., Jenni, O. G., Molinari, L., and Largo, R. H. Sleep duration from infancy to adolescence: Reference values and generational trends. *Pediatrics* 111:302–7 (2003).

Iwaniuk, A. N., Nelson, J. E., and Pellis, S. M. Do big-brained animals play more? Comparative analyses of play and relative brain size in mammals. *Journal of Comparative Psychology* 115:29–41 (2001).

Jablonka, E., and Raz, G. Transgenerational epigenetic inheritance: Prevalence, mechanisms, and implications for the study of heredity and evolution. *Quarterly Review of Biology* 84:131–76 (2009).

Jacobson, J. W., Foxx, R. M., and Mulick, J. A., eds. *Controversial Therapies for Developmental Disabilities: Fad, Fashion, and Science in Professional Practice*. Mahwah, NJ: Psychology Press, 2005.

Jaeggi, S. M., Buschkuehl, M., Jonides, J., and Perrig, W. J. Improving fluid intelligence with training on working memory. *Proceedings of the National Academy of Sciences U.S.A.* 105:6829–33 (2008).

Jaffee, S., and Hyde, J. S. Gender differences in moral orientation: A meta-analysis. *Psychological Bulletin* 126:703–26 (2000).

Jain, A., Concato, J., and Leventhal, J. M. How good is the evidence linking breastfeeding and intelligence? *Pediatrics* 109:1044–53 (2002).

Jensen, J. K., and Neff, D. L. Development of basic auditory discrimination in preschool children. *Psychological Science* 4:104–7 (1993).

Joëlsa, M., Pua, Z., Wiegerta, O., Oitzlb, M. S., and Kruger, H. J. Learning under stress: How does it work? *Trends in Cognitive Sciences* 10:152–58 (2006).

Johnson, J. S., and Newport, E. L. Critical period effects in second language learning: The influence of maturational state on the acquisition of English as a second language. *Cognitive Psychology* 21:60–99 (1989).

Johnson, M. H., Dziurawiec, S., Eills, H., and Morton, J. Newborn's preferential tracking of face-like stimuli and its subsequent decline. *Cognition* 40:1–19 (1991).

Jones, J. L., Anderson, B., Schulkin, J., Parise, M. E., and Eberhard, M. L. Sushi in pregnancy, parasitic diseases—obstetrician survey. *Zoonoses and Public Health.* doi: 10.1111/j.1863-2378.2009.01310.x (2009).

Jones, L. A., Sinnott, L. T., Mutti, D. O., Mitchell, G. L., Moeschberger, M. L., and Zadnik, K. Parental history of myopia, sports and outdoor activities, and future myopia. *Investigative Ophthalmology and Visual Science* 48:3524–32 (2007).

Kagen, J. Biology, context, and developmental inquiry. *Annual Review of Psychology* 54:1–23 (2003).

Kaitz, M., Good, A., Rokem, A. M., and Eidelman, A. I. Mothers' recognition of their newborns by olfactory cues. *Developmental Psychobiology* 20:587–91 (1987).

Kandel, E. R., Schwartz, J. H., and Jessell, T. M. *Principles of Neural Science.* 4th edition. New York: McGraw-Hill, 2000.

Karmel, B. Z., Gardner, J. M., Meade, L. S., Cohen, I. L., London, E., Flory, M. J., Lennon, E. M., Miroshnichenko, I., Rabinowitz, S., Parab, S., Barone, A., and Harin, A. Early medical and behavioral characteristics of NICU infants later classified with ASD. *Pediatrics* 126:457–67 (2010).

Kavšek, M. Predicting later IQ from infant visual habituation and dishabituation: A meta-analysis. *Applied Developmental Psychology* 25:369–93 (2004).

Kawachi, I., Adler, N. E., and Dow, W. H. Money, schooling, and health: Mechanisms and causal evidence. *Annals of the New York Academy of Sciences* 1186:56–68 (2010).

Kazdin, A. E., and Rotella, C. *The Kazdin Method for Parenting the Defiant Child: With No Pills, No Therapy, No Contest of Wills.* New York: Houghton Mifflin Harcourt, 2009.

Kei, J., McPherson, B., Smyth, V., Latham, S., and Loscher, J. Transient evoked otoacoustic emissions in infants: Effects of gender, ear asymmetry and activity status. *Audiology* 36:61–71 (1997).

Kelly, D., Quinn, P., Slater, A., Lee, K., Ge, L., and Pascalis, O. The other-race effect develops during infancy: Evidence of perceptual narrowing. *Psychological Science* 18:1084–87 (2007).

Kessel, C., and Linn, M. C. Grades or

scores: Predicting future college mathematics performance. *Educational Measurement: Issues and Practice* 15:10–14 (1996).

Keuroghlian, A. S., and Knudsen, E. I. Adaptive auditory plasticity in developing and adult animals. *Progress in Neurobiology* 82:109–21 (2007).

Khashan, A. S., Abel, K. M., McNamee, R., Pedersen, M. G., Webb, R. T., Baker, P. N., Kenny, L. C., and Mortensen, P. B. Higher risk of offspring schizophrenia following antenatal maternal exposure to severe adverse life events. *Archives of General Psychiatry* 65:146–52 (2008).

Kimura, D. *Sex and Cognition*. Cambridge, MA: MIT Press, 2000.

Kinney, D. K., Miller, A. M., Crowley, D. J., Huang, E., and Gerber, E. Autism prevalence following prenatal exposure to hurricanes and tropical storms in Louisiana. *Journal of Autism and Developmental Disorders* 38:481–88 (2008).

Kling, K. C., Hyde, J. S., Showers, C. J., and Buswell, B. N. Gender differences in self-esteem: A meta-analysis. *Psychological Bulletin* 125:470–500 (1999).

Knops, A., Thirion, B., Hubbard, E. M., Michel, V., and Dehaene, S. Recruitment of an area involved in eye movements during mental arithmetic. *Science* 324:1583–85 (2009).

Knudsen, E. I. Sensitive periods in the development of brain and behavior. *Journal of Cognitive Neuroscience* 16:1412–25 (2004).

Knudsen, E. I., Heckman, J. J., Cameron, J. L., and Shonkoff, J. P. Economic, neurobiological, and behavioral perspectives on building America's future workforce. *Proceedings of the National Academy of Sciences* 103:10155–62 (2006).

Kochanska, G., and Aksan, N. Children's conscience and self-regulation. *Journal of Personality* 74:1587–1618 (2006).

Kochanska, G., Philibert, R. A., and Barry, R. A. Interplay of genes and early mother-child relationship in the development of self-regulation from toddler to preschool age. *Journal of Child Psychology and Psychiatry* 50:1331–38 (2009).

Kovács, Å. M., and Mehler, J. Flexible learning of multiple speech structures in bilingual infants. *Science* 325:611–12 (2009).

Kovas, Y., Hayiou-Thomas, M. E., Oliver, B., Dale, P. S., Bishop, D. V., and Plomin, R. Genetic influences in different aspects of language development: The etiology of language skills in 4.5-year-old twins. *Child Development* 76:632–51 (2005).

Kramer, M. S., et al. Breastfeeding and child cognitive development: New evidence from a large randomized trial. *Archives of General Psychiatry* 65:578–84 (2008).

Kuhl, P., and Rivera-Gaxiola, M. Neural substrates of language acquisition. *Annual Review of Neuroscience* 31:511–34 (2008).

Laland, K. N., Odling-Smee, J., and Myles, S. How culture shaped the human genome: Bringing genetics and the human sciences together. *Nature Reviews Genetics* 11:137–48 (2010).

LaPlante, D. P., Brunet, A., Schmitz, N., Ciampi, A., and King, S. Project ice storm: Prenatal maternal stress affects cognitive and linguistic functioning in 5 ½-year-old children. *Journal of the American Academy of Child and Adolescent Psychiatry* 47:1063–72 (2008).

Lee, K., Lim, Z. Y., Yeong, S. H. M., Ng, S. F., Venkatraman, V., and Chee, M. W. L. Strategic differences in algebraic problem solving: Neuroanatomical correlates. *Brain Research* 1155:163–71 (2007).

Lenroot, R. K., Gogtay, N., Greenstein, D. K., Wells, E. M., Wallace, G. L., Clasen, L. S., Blumenthal, J. D., Lerch, J., Zijdenbos, A. P., Evans, A. C., Thompson, P. M., and Giedd, J. N. Sexual dimorphism of brain developmental trajectories during childhood and adolescence. *NeuroImage* 36:1065–73 (2007).

Leonard, D. K., and Jiang, J. Gender bias and the college predictions of the SATs: A cry of despair. *Research in Higher Education* 40:375–407 (1999).

Leppänen, J. M., and Nelson, C. A. Tuning the developing brain to social signals of emotions. *Nature Reviews Neuroscience* 10:37–47 (2009).

Leuchter, A. F. Review of *Healing the Hardware of the Soul* by Daniel Amen. *American Journal of Psychiatry* 166:625 (2009).

Levi, D. M. Perceptual learning in adults with amblyopia: A reevaluation of critical periods in human vision. *Developmental Psychobiology* 46:222–32 (2005).

Levine, S. C., Vasilyeva, M., Lourenco, S. F., Newcombe, N. S., and Huttenlocher, J. Socioeconomic status modifies the sex difference in spatial skill. *Psychological Science* 16:841–45 (2005).

Lewis, T. L., and Maurer, D. Effects of early pattern deprivation on visual development. *Optometry and Vision Science* 86:640–46 (2009).

Li, R., Polat, U., Makous, W., and Bavelier, D. Enhancing the contrast sensitivity function through action video game training. *Nature Neuroscience* 12:549–51 (2009).

Limperopoulos, C. Extreme prematurity, cerebellar injury, and autism. *Seminars in Pediatric Neurology* 17:25–29 (2010).

Linebarger, D. L., and Walker, D. Infants' and toddlers' television viewing and language outcomes. *American Behavioral Scientist* 48:624–45 (2005).

Linn, M. C., and Kessel, C. Success in mathematics: Increasing talent and gender diversity among college majors. *Issues in Mathematics Education* 6:101–44 (1996).

Löken, L. S., Wessberg, J., Morrison, I., McGlone, F., and Olausson, H. Coding of pleasant touch by unmyelinated afferents in humans. *Nature Neuroscience* 12:547–48 (2009).

Luciana, M. Adolescent brain development: Current themes and future directions. Introduction to the special issue. *Brain and Cognition* 72:1–5 (2010).

Maccoby, E. E. Parenting and its effects on children: On reading and misreading behavior genetics. *Annual Review of Psychology* 51:1–27 (2000).

———.*The Two Sexes: Growing Up Apart, Coming Together.* Cambridge, MA: Harvard University Press, 1998.

Maccoby, E. E., and Jacklin, C. N. Gender segregation in childhood. *Advances in Child Development and Behavior* 20:239–87 (1987).

Mannuzza, S., Klein, R. G., Bonagura, N., Malloy, P., Giampino, T. L., and Addalli, K. A. Hyperactive boys almost grown up: V. Replication of psychiatric status. *Archives of General Psychiatry* 48:77–83 (1991).

Mannuzza, S., Klein, R. G., Truong, N. L., Moulton, J. L., III, Roizen, E. R., and

Howell, K. H. Age of methylphenidate treatment initiation in children with ADHD and later substance abuse: Prospective follow-up into adulthood. *American Journal of Psychiatry* 165:604–9 (2008).

Marois, R., and Ivanoff, J. Capacity limits of information processing in the brain. *Trends in Cognitive Sciences* 9:296–305 (2005).

Mather, J. A., and Anderson, R. C. Exploration, play and habituation in octopuses (Octopus dofleini). *Journal of Comparative Psychology* 113:333–38 (1999).

Matthews, K. A., Gallo, L. C., and Taylor, S. E. Are psychosocial factors mediators of socioeconomic status and health connections? A progress report and blueprint for the future. *Annals of the New York Academy of Sciences* 1186:146–73 (2010).

McAdams, D. P., and Olson, B. D. Personality development: Continuity and change over the life course. *Annual Review of Psychology* 61:517–42 (2010).

McCarthy, E. S. Greek and Roman lore of animal-nursed infants. Papers of the Michigan Academy of Science, Arts, and Letters 4:15–40 (1925).

McClure, E. B. A meta-analytic review of sex differences in facial expression processing and their development in infants, children and adolescents. *Psychological Bulletin* 126:424–53 (2000).

McEwen, B. S., and Gianaros, P. J. Central role of the brain in stress and adaptation: Links to socioeconomic status, health, and disease. *Annals of the New York Academy of Sciences* 1186:190–222 (2010).

McGlone, M. S., and Aronson, J. Stereotype threat, identity salience, and spatial reasoning. *Journal of Applied Developmental Psychology* 27:486–93 (2006).

McKee, A. C., Cantu, R. C., Nowinski, C. J., Hedley-Whyte, E. T., Gavett, B. E., Budson, A. E., Santini, V. E., Lee, H. S., Kubilus, C. A., and Stern, R. A. Chronic traumatic encephalopathy in athletes: Progressive tauopathy after repetitive head injury. *Journal of Neuropathology and Experimental Neurology* 68:709–35 (2009).

Mednick, S., and Ehrman, M. *Take a Nap! Change Your Life.* New York: Workman, 2006.

Mednick, S. C., Nakayama, K. Cantero, J. L., Atienza, M., Levin, A. A., Pathak, N., and Stickgold, R. The restorative effect of naps on perceptual deterioration. *Nature Neuroscience* 5:677–81 (2002).

Mehler, J., and Bever, T. G. Cognitive capacity of very young children. *Science* 158:141–42 (1967).

Mehler, J., Jusczyk, P., Lambertz, G., Halsted, N., Bertoncini, J., and Amiel-Tison, C. A. Precursor of language acquisition in young infants. *Cognition* 29:143–78 (1988).

Meisami, E., and Timiras, P. S., eds. *Handbook of Human Growth and Developmental Biology*. Vol. I, part B. Boca Raton, FL: CRC Press, 1988.

Mennella, J. A., and Beauchamp, G. K. Flavor experiences during formula feeding are related to preferences during childhood. *Early Human Development* 68:71–82 (2002).

Mennella, J. A., and Gerrish, C. J. Effects of exposure to alcohol in mother's milk on infant sleep. *Pediatrics*, 101:e2 (1998).

Mennella, J. A., Jagnow, C. P., and

Beauchamp, G. K. Prenatal and postnatal flavor learning by human infants. *Pediatrics* 107:e88 (2001).

Mesman, J., van IJzendoorn, M. H., and Bakermans-Kranenburg, M. J. The many faces of the Still-Face Paradigm: A review and meta-analysis. *Developmental Review* 29:120–62 (2009).

Milner, J. A., and Allison, R. G. The role of dietary fat in child nutrition and development: Summary of an ASNS workshop. *Journal of Nutrition* 129:2094–2105 (1999).

Mink, J. W. The basal ganglia: Focused selection and inhibition of competing motor programs. *Progress in Neurobiology* 50:381–425 (1996).

Mistretta, C. M., and Bradley, R. M. Taste and swallowing in utero: A discussion of fetal sensory function. *British Medical Bulletin* 31:80–84 (1984).

Mondloch, C. J., Lewis, T. L., Budreau, D. R., Maurer, D., Dannemiller, J. D., Stephens, B. R., et al. Face perception during early infancy. *Psychological Science* 10:419–22 (1999).

Moore, D. S., and Johnson, S. P. Mental rotation in human infants. *Psychological Science* 19:1063–66 (2008).

Moseley, M. J., Fielder, A. R., and Stewart, C. E. The optical treatment of amblyopia. *Optometry and Visual Science* 86:629–33 (2009).

Moster, D., Lie, R. T., and Markestad, T. Long-term medical and social consequences of preterm birth. *New England Journal of Medicine* 359:262–73 (2008).

Müller, K., Ebner, B., and Hömberg, V. Maturation of fastest afferent and efferent central and peripheral pathways: No evidence for a constancy of central conduction delays. *Neuroscience Letters* 166:9–12 (1994).

Nakata, T., Trehub, S. E., Mitani, C., and Kanda, Y. Pitch and timing in the songs of deaf children with cochlear implants. *Music Perception* 24:147–54 (2006).

Nandi, R., and Luxon, L. M. Development and assessment of the vestibular system. *International Journal of Audiology* 47:566–77 (2008).

Nassar, N., Dixon, G., Bourke, J., Bower, C., Glasson, E., de Klerk, N., and Leonard, H. Autism spectrum disorders in young children: Effect of changes in diagnostic practices. *International Journal of Epidemiology* 38:1245–54 (2009).

Neumark-Sztainer, D. R., Wall, M. M., Haines, J. I., Story, M. T., Sherwood, N. E., and van den Berg, P. A. Shared risk and protective factors for overweight and disordered eating in adolescents. *American Journal of Preventive Medicine* 33:359–69 (2007).

Newport, E. L., Bavelier, D., and Neville, H. J. Critical thinking about critical periods: Perspectives on a critical period for language acquisition. In *Language, Brain, and Cognitive Development: Essays in Honor of Jacques Mehler*, ed. E. Dupoux, 481–502. Cambridge, MA: MIT Press, 2001.

Nicklas, T. A., Webber, L. S., Koschak, M. L., and Berenson, G. S. Nutrient adequacy of low fat intakes for children: The Bogalusa heart study. *Pediatrics* 89:221–28 (1992).

Nieder, A., and Dehaene, S. Representation of number in the brain. *Annual Review of Neuroscience* 32:185–208 (2009).

Nir, Y., and Tononi, G. Dreaming and the brain: From phenomenology to neurophysiology. *Trends in Cognitive Sciences* 14:88–100 (2009).

Okagaki, L., and Frensch, P. A. Effects of video game playing on measures of spatial performance: Gender effects in late adolescence. *Journal of Applied Developmental Psychology* 15:33–58 (1994).

Oken, E., Radesky, J. S., Wright, R. O., Bellinger, D. C., Amarasiriwardena, C. J., Kleinman, K. P., Hu, H., and Gillman, M. W. Maternal fish intake during pregnancy, blood mercury levels, and child cognition at age 3 years in a US cohort. *American Journal of Epidemiology* 167:1171–81 (2008).

Olausson, H., Lamarre, Y., Backlund, H., Morin, C., Wallin, B. G., Starck, G., Ekholm, S., Strigo, I., Worsley, K., Vallbo, Å. B., and Bushnell, M. C. Unmyelinated tactile afferents signal touch and project to insular cortex. *Nature Neuroscience* 5:900–904 (2002).

Ophir, E., Nass, C., and Wagner, A. D. Cognitive control in media multitaskers. *Proceedings of the National Academy of Sciences USA* 106:15583–87 (2009).

Ozel, S., Larue, J., and Molinaro, C. Relation between sport and spatial imagery: Comparison of three groups of participants. *Journal of Psychology* 138:49–63 (2004).

Palmen, S. J. M. C., van Engeland, H., Hof, P. R., and Schmitz, C. Neuropathological findings in autism. *Brain* 127:2572–83 (2004).

Parlade, M. V., Messinger, D. S., Delgado, C. E. F., Kaiser, M. Y., Van Hecke, A. V., and Mundy, P. C. Anticipatory smiling: Linking early affective communication and social outcome. *Infant Behavior and Development* 32:33–43 (2009).

Pasterski, V., Hindmarsh, P., Geffner, M., Brook, C., Brain, C., and Hines, M. Increased aggression and activity level in 3- to 11-year-old girls with congenital adrenal hyperplasia (CAH). *Hormones and Behavior* 52:368–74 (2007).

Peake, P. K., Hebl, M., and Mischel, W. Strategic attention deployment for delay of gratification in working and waiting situations. *Developmental Psychology* 38:313–26 (2002).

Petersen, J. L., and Hyde, J. S. A meta-analytic review of research on gender differences in sexuality: 1993 to 2007. *Psychological Bulletin* 136:21–38 (2010).

Pica, P., Lemer, C., Izard, V., and Dehaene, S. Exact and approximate arithmetic in an Amazonian indigene group. *Science* 306:499–503 (2004).

Plomin, R., Asbury, K., Dip, P. G., and Dunn, J. Why are children in the same family so different? Nonshared environment a decade later. *Canadian Journal of Psychiatry* 46:225–33 (2001).

Pontifex, M. B., Raine, L. B., Johnson, C. R., Chaddock, L., Voss, M. W., Cohen, N. J., Kramer, A. F., and Hillman, C. H. Cardiorespiratory fitness and the flexible modulation of cognitive control in preadolescent children. *Journal of Cognitive Neuroscience* 23:1332–45 (2011).

Posner, M. I., and Rothbart, M. K. Research on attention networks as a model for the integration of psychological science. *Annual Review of Psychology* 58:1–23 (2007).

Power, T. G. Stress and coping in childhood: The parents' role. *Parenting: Science and Practice* 4:271–317 (2004).

Preston, S. D., and de Waal, F. B. M. Empathy: Its ultimate and proximate bases. *Behavioral and Brain Sciences* 25:1–72 (2002).

Quinn, P. C. Beyond prototypes: Asymmetries in infant categorization and what they teach us about the mechanisms guiding early knowledge acquisition. *Advances in Child Development and Behavior* 29:161–93 (2002).

Quinn, P. C., and Liben, L. S. A sex difference in mental rotation in young infants. *Psychological Science* 19:1067–70 (2008).

Quirk, G. J., and Mueller, D. Neural mechanisms of extinction learning and retrieval. *Neuropsychopharmacology* 33:56–72 (2008).

Rakic, P., Bourgeois, J.-P., and Goldman-Rakic, P. S. Synaptic development of the cerebral cortex: Implications for learning, memory, and mental illness. *Progress in Brain Research* 102:227–43 (1994).

Ramus, F., Rosen, S., Dakin, S. C., Day, B. L., Castellote, J. M., White, S., and Frith, U. Theories of developmental dyslexia: Insights from a multiple case study of dyslexic adults. *Brain* 126:841–65 (2003).

Redgrave, P., Prescott, T. J., and Gurney, K. Is the short-latency dopamine response too short to signal reward error? *Trends in Neuroscience* 22:146–51 (1999).

Reid, V. M., and Striano, T. Adult gaze influences infant attention and object processing: Implications for cognitive neuroscience. *European Journal of Neuroscience* 21:1763–66 (2005).

Reiss, D., and Leve, L. D. Genetic expression outside the skin: Clues to mechanisms of genotype × environment interaction. *Development and Psychopathology* 19:1005–27 (2007).

Rice, G. E., and Gainer, P. "Altruism" in the albino rat. *Journal of Comparative and Physiological Psychology* 55:123–25 (1962).

Rizzolatti, G., and Sinigaglia, C. The functional role of the parieto-frontal mirror circuit: Interpretations and misinterpretations. *Nature Reviews Neuroscience* 11:264–74 (2010).

Roffwarg, H. P., Muzio, J. N., and Dement, W. C. Ontogenetic development of the human sleep-dream cycle. *Science* 152:604–19 (1966).

Rohrer, D., and Pashler, H. Recent research on human learning challenges conventional instructional strategies. *Educational Researcher* 39:406–12 (2010).

Rolls, E. T. Brain mechanisms underlying flavour and appetite. *Philosophical Transactions of the Royal Society B* 361:1123–36 (2006).

Romeo, R. D., and McEwen, B. S. Stress and the adolescent brain. *Annals of the New York Academy of Sciences* 1094:202–14 (2006).

Rose, K., Smith, W., Morgan, I., and Mitchell, P. The increasing prevalence of myopia: Implications for Australia. *Clinical and Experimental Ophthalmology* 29:116–20 (2001).

Rose, K. A., Morgan, I. G., Ip, J., Kifley, A., Huynh, S., Smith, W., and Mitchell, P. Outdoor activity reduces the prevalence of myopia in children. *Ophthalmology* 115:1279–85 (2008a).

Rose, K. A., Morgan, I. G., Smith, W., Burlutsky, G., Mitchell, P., and Saw, S.-M. Myopia, lifestyle, and schooling in students of Chinese ethnicity in Singapore and Sydney. *Archives of Ophthalmology* 126:527–30 (2008b).

Rothbart, M. K., Ahadi, S. A., and Evans, D. E. Temperament and personality: Origins and outcomes. *Journal of Personality and Social Psychology* 78:122–35 (2000).

Rovee-Collier, C., and Barr, R. Infant

learning and memory. In *Blackwell Handbook of Infant Development*, ed. G. Bremner and A. Fogel, 139–68. Malden, MA: Blackwell, 2001.

Rowe, D. C., and Rodgers, J. L. Poverty and behavior: Are environmental measures nature and nurture? *Developmental Review* 17:358–75 (1997).

Rubin, K. H., Bukowski, W., and Parker, J. G. Peer interactions, relationships, and groups. In *Handbook of Child Psychology*, vol. 3, *Social, Emotional, and Personality Development*, ed. N. Eisenberg, 571–645. New York: Wiley, 2006.

Rubin, K. H., Coplan, R. J., and Bowker, J. C. Social withdrawal in childhood. *Annual Review of Psychology* 60:141–71 (2009).

Ruffman, T., Perner, J., Naito, M., Parkin, L., and Clements, W. A. Older (but not younger) siblings facilitate false belief understanding. *Developmental Psychology* 34:161–74 (1998).

Rutter, M. Gene-environment interdependence. *Developmental Science* 10:12–18 (2007).

Saffran, J. R., Aslin, R. N., and Newport, E. L. Statistical learning by 8-month-old infants. *Science* 274:1926–28 (1996).

Sallows, G. O., and Graupner, T. D. Intensive behavioral treament for children with autism: Four-year outcome and predictors. *American Journal of Mental Retardation* 110:417–38 (2005).

Salomão, S. R., and Ventura, D. F. Large sample population age norms for visual acuities obtained with Vistech–Teller acuity cards. *Investigative Ophthalmology and Visual Science* 36:657–70 (1995).

Sanes, D. H., Reh, T. A., and Harris, W. A. *Development of the Nervous System*. 2nd edition. New York: Academic Press, 2005.

Sapolsky, R. M. The influence of social hierarchy on primate health. *Science* 308:648–52 (2005).

———. *Why Zebras Don't Get Ulcers*. 3rd edition. New York: Holt, 2004.

Saxe, R., Carey, S., and Kanwisher, N. Understanding other minds: Linking developmental psychology and functional neuroimaging. *Annual Review of Psychology* 55:87–124 (2004).

Schaal, B. Olfaction in infants and children: Developmental and functional perspectives. *Chemical Senses* 13:145–90 (1988).

Schellenberg, E. G. Music and cognitive abilities. *Current Directions in Psychological Science* 14:317–20 (2005).

Schlaug, G., Jäncke, L., Huang, Y., Staiger, J. F., and Steinmetz, H. Increased corpus callosum size in musicians. *Neuropsychologia* 33:1047–55 (1995).

Schneider, P., Scherg, M., Dosch, H. G., Specht, H. J., Gutschalk, A., and Rupp, A. Morphology of Heschl's gyrus reflects enhanced activation in the auditory cortex of musicians. *Nature Neuroscience* 5:688–94 (2002).

Servin, A., Bohlin, G., and Berlin, L. Sex differences in 1-, 3-, and 5-year-olds' toy-choice in a structured play-session. *Scandinavian Journal of Psychology*, 40:43–48 (1999).

Shaw, P. Intelligence and the developing human brain. *BioEssays* 29:962–73 (2007).

Shaw, P., Eckstrand, K., Sharp, W., Blumenthal, J., Lerch, J. P., Greenstein, D., Clasen, L., Evans, A., Giedd, J., and Rapoport, J. L. Attention-deficit/hyperactivity disorder is characterized by a delay in cortical maturation. *Proceedings of the National Academy of Sciences USA* 104:19649–54 (2007).

Shaw, P., Greenstein, D., Lerch, J., Clasen, L., Lenroot, R., Gogtay, N., Evans, A., Rapoport, J., and Giedd, J. Intellectual ability and cortical development in children and adolescents. *Nature* 440:676–79 (2006a).

Shaw, P., Lerch, J., Greenstein, D., Sharp, W., Clasen, L., Evans, A., Giedd, J., Castellanos, F. X., and Rapoport, J. Longitudinal mapping of cortical thickness and clinical outcome in children and adolescents with attention-deficit/hyperactivity disorder. *Archives of General Psychiatry* 63:540–49 (2006b).

Sheese, B. E., Voelker, P. M., Rothbart, M. K., and Posner, M. I. Parenting quality interacts with genetic variation in dopamine receptor D4 to influence temperament in early childhood. *Development and Psychopathology* 19:1039–46 (2007).

Shoda, Y., Mischel, W., and Peake, P. K. Predicting adolescent cognitive and self-regulatory competencies from preschool delay of gratification: Identifying diagnostic conditions. *Developmental Psychology* 26:978–86 (1990).

Shorter, E. *Before Prozac: The Troubled History of Mood Disorders in Psychiatry.* New York: Oxford University Press, 2008.

Sibley, B. A., and Etnier, J. L. The relationship between physical activity and cognition in children: A meta-analysis. *Pediatric Exercise Science* 15:243–56 (2003).

Singer, T. The neuronal basis and ontogeny of empathy and mind reading: Review of literature and implication for future research. *Neuroscience and Biobehavioral Reviews* 30:855–63 (2006).

Siok, W. T., Niu, Z., Jin, Z., Perfetti, C. A., and Tan, L. H. A structural-functional basis for dyslexia in the cortex of Chinese readers. *Proceedings of the National Academy of Sciences USA* 105:5561–66 (2008).

Sisk, S., and Foster, D. L. The neural basis of puberty and adolescence. *Nature Neuroscience* 7:1040–47 (2004).

Skinner, E. A., and Zimmer-Gembeck, M. J. The development of coping. *Annual Review of Psychology* 58:119–44 (2007).

Smetana, J. G., Campione-Barr, N., and Metzger, A. Adolescent development in interpersonal and societal contexts. *Annual Review of Psychology* 57:255–84 (2006).

Snowdon, D. A., Ostwald, S. K., Kane, R. L., and Keenan, N. L. Years of life with good and poor mental and physical function in the elderly. *Journal of Clinical Epidemiology* 42:1055–66 (1989).

Sohn, M.-H., Goode, A., Koedinger, K. R., Stenger, V. A., Fissell, K., Carter, C. S., and Anderson, J. R. Behavioral equivalence, but not neural equivalence—neural evidence of alternative strategies in mathematical thinking. *Nature Neuroscience* 7:1193–94 (2004).

Spear, L. P. *The Behavioral Neuroscience of Adolescence.* New York: W. W. Norton, 2010.

Spelke, E. S., and Kinzler, K. D. Core knowledge. *Developmental Science* 10:89–96 (2007).

Squire, L. R. *Memory and Brain.* New York: Oxford University Press, 1987.

Squire, L. R., and Zola-Morgan, S. M. The medial temporal lobe memory system. *Science* 253:1380–85 (1991).

Steinberg, L. A behavioral scientist looks at the science of adolescent brain development. *Brain and Cognition* 72:160–64 (2010).

Strand, P. S. A modern behavioral perspective on child conduct disorder: Integrating behavioral momentum and matching theory. *Clinical Psychology Review* 20:593–615 (2000).

Subrahmanyam, K., and Greenfield, P. M. Effect of video game practice on spatial skills in girls and boys. *Journal of Applied Developmental Psychology* 15:13–32 (1994).

Sullivan, R. M. Review: Olfaction in the human infant. *Sense of Smell Institute White Paper* (July 2000). http://www.senseofsmell.org/papers/Sullivan_olfaction_baby.pdf.

Suomi, S. J. Early determinants of behaviour: Evidence from primate studies. *British Medical Bulletin* 53:170–84 (1997).

———. Risk, resilience, and gene × environment interactions in rhesus monkeys. *Annals of the New York Academy of Sciences* 1994:52–62 (2006).

Tamis-LeMonda, C. S., Bornstein, M. H., Cyphers, L., Toda, S., and Ogino, M. Language and play at one year: A comparison of toddlers and mothers in the United States and Japan. *International Journal of Behavioral Development* 15:19–42 (1992).

Tan, L. H., Spinks, J. A., Eden, G. F., Perfetti, C. A., and Siok, W. T. Reading depends on writing, in Chinese. *Proceedings of the National Academy of Sciences USA* 102:8781–85 (2005).

Tang, Y.-Y., and Posner, M. I. Attention training and attention state training. *Trends in Cognitive Sciences* 13:222–27 (2009).

Tau, G. Z., and Peterson, B. S. Normal development of brain circuits. *Neuropsychopharmacology* 35:147–68 (2010).

Taubes, G. Do we really know what makes us healthy? *New York Times Sunday Magazine*, September 16, 2007.

Thomas, G. E. Student and institutional characteristics as determinants of the prompt and subsequent four-year college graduation of race and sex groups. *Sociological Quarterly* 22:327–45 (1981).

Thompson, B. L., Levitt, P., and Stanwood, G. D. Prenatal exposure to drugs: Effects on brain development and implications for policy and education. *Nature Reviews Neuroscience* 10:303–12 (2009).

Townsend, F. Taking "Born to Rebel" seriously: The need for independent review. *Politics and the Life Sciences* 19:205–10 (2000).

Trainor, L. J., Tsang, C. D., and Cheung, V. H. W. Preference for sensory consonance in 2- and 4-month-old infants. *Music Perception* 20:187–94 (2002).

Tripp, G., and Wickens, J. R. Neurobiology of ADHD. *Neuropharmacology* 57:579–89 (2009).

Turkeltaub, P. E., Gareau, L., Flowers, D. L., Zeffiro, T. A., and Eden, G. F. Development of neural mechanisms for reading. *Nature Neuroscience* 6:767–73 (2003).

Turkheimer, E., Haley, A., Waldron, M., D'Onofrio, B., and Gottesman, I. I. Socioeconomic status modified heritability of IQ in young children. *Psychological Science* 14:623–28 (2003).

Uher, R., and McGuffin, P. The moderation by the serotonin transporter gene of environmental adversity in the etiology of depression: 2009 update. *Molecular Psychiatry* 15:18–22 (2010).

van IJzendoorn, M. H., Juffer, F., and Klein Poelhuis, C. W. Adoption and cognitive development: A meta-analytic comparison of adopted and nonadopted children's IQ and school performance. *Psychological Bulletin* 131:301–16 (2005).

Vanderschuren, L. J. M. J., Niesink, R. J. M., and van Ree, J. M. The neurobiology of social play behavior in rats. *Neuroscience and Biobehavioral Reviews* 21:309–26 (1997).

Volkmar, F., Chawarska, K., and Klin, A. Autism in infancy and early childhood. *Annual Review of Psychology* 56:315–36 (2005).

Voyer, D., Voyer, S., and Bryden, M. P. Magnitude of sex differences in spatial abilities: A meta-analysis and consideration of critical variables. *Psychological Bulletin* 117:250–70 (1995).

Wager, T. D., Phan, K. L., Liberzon, I., and Taylor, S. F. Valence, gender, and lateralization of functional brain anatomy in emotion: A meta-analysis of findings from neuroimaging. *NeuroImage* 19:513–31 (2003).

Wagner, R. K., Torgesen, J. K., Rashotte, C. A., Hecht, S. A., Barker, T. A., Burgess, S. R., Donahue, J., and Garon, T. Changing relations between phonological processing abilities and word-level reading as children develop from beginning to skilled readers: A 5-year longitudinal study. *Developmental Psychology* 33:468–79 (1997).

Wallen, K. Hormonal influences on sexually differentiated behavior in nonhuman primates. *Frontiers in Neuroendocrinology* 26:7–26 (2005).

Wang, R. F., Hermer, L., and Spelke, E. S. Mechanisms of reorientation and object localization by children: A comparison with rats. *Behavioral Neuroscience* 113:475–85 (1999).

Wang, S.-H., and Morris, R. G. M. Hippocampal-neocortical interactions in memory formation, consolidation, and reconsolidation. *Annual Review of Psychology* 61:49–79 (2010).

Wazana, A., Bresnahan, M., and Kline, J. The autism epidemic: Fact or artifact? *Journal of the American Academy of Child and Adolescent Psychiatry* 46:721–30 (2007).

Webster, R. *Why Freud Was Wrong: Sin, Science, and Psychoanalysis.* New York: Basic Books, 1995.

Weissbluth, M. *Healthy Sleep Habits, Happy Child.* 3rd edition. New York: Ballantine Books, 2003.

Welsh, D. K., Takahashi, J. S., and Kay, S. A. Suprachiasmatic nucleus: Cell autonomy and network properties. *Annual Review of Physiology* 72:551–77 (2010).

Werker, J. F., and Byers-Heinlein, K. Bilingualism in infancy: First steps in perception and comprehension. *Trends in Cognitive Sciences* 12:144–51 (2008).

Wheeler, S. C., and Petty, R. E. The effects of stereotype activation on behavior: A review of possible mechanisms. *Psychological Bulletin* 127:797–826 (2001).

Whitehurst, G. J., Falco, F. L., Lonigan, C. J., Fischel, J. E., Debaryshe, D. B., Valdez-Menchaca, M. C., and Caulfield, M. Accelerating language development through picture book reading. *Developmental Psychology* 40:395–403 (1988).

Willey, L. H. *Pretending to Be Normal: Living with Asperger's Syndrome.* London: Jessica Kingsley, 1999.

Williams, P. Cultural impressions of the wolf, with specific reference to the man-eating wolf in England. University of Sheffield, PhD thesis (2003).

Winberg, J., and Porter, R. H. Olfaction and human neonatal behaviour: Clinical implications. *Acta Paediatrica* 87:6–10 (1998).

Witt, M., and Reutter, K. Embryonic and early fetal development of human taste buds: A transmission electron microscopical study. *Anatomical Record* 246:507–23 (1996).

Witter, F. R., Zimmerman, A. W., Reichmann, J. P., and Connors, S. L. In utero beta 2 adrenergic agonist exposure and adverse neurophysiologic and behavioral outcomes. *American Journal of Obstetrics and Gynecology* 201:553–59 (2009).

Wittmann, M., Dinich, J., Merrow, M., and Roenneberg, T. Social jetlag: Misalignment of biological and social time. *Chronobiology International* 23:497–509 (2006).

Wolff, U., and Lundberg, I. The prevalence of dyslexia among art students. *Dyslexia* 8:34–42 (2002).

Wong, C. C. Y., Caspi, A., Williams, B., Craig, I. W., Houts, R., Ambler, A., Moffitt, T. E., and Mill, J. A longitudinal study of epigenetic variation in twins. *Epigenetics* 5:1–11 (2010).

Woodward, A. L. Infants selectively encode the goal object of an actor's reach. *Cognition* 69:1–34 (1998).

Yarmolinsky, D. A., Zuker, C. S., and Ryba, N. J. Common sense about taste: From mammals to insects. *Cell* 139:234-44 (2009).

Yin, W. G., and Weekes, B. S. Dyslexia in Chinese: Clues from cognitive neuropsychology. *Annals of Dyslexia* 53:255–79 (2003).

Zatorre, R. J., Chen, J. L., and Penhune, V. B. When the brain plays music: Auditory-motor interactions in music perception and production. *Nature Reviews Neuroscience* 8:547–58 (2007).

Zhang, T.-Y., and Meaney, M. J. Epigenetics and the environmental regulation of the genome and its function. *Annual Review of Psychology* 61:439–66 (2010).

Zhao, G. Q., Zhang, Y., Hoon, M. A., Chandrashekar, J., Erlenbach, I., Ryba, N. J., and Zuker, C. S. The receptors for mammalian sweet and umami taste. *Cell* 115:255-66 (2003).

INDEX

Note: *Italic* page numbers refer to illustrations.
Note: **Bold** terms appear in the glossary.